EPIC
EGGS

EPIC EGGS

THE POULTRY ENTHUSIAST'S
COMPLETE AND ESSENTIAL GUIDE
TO THE MOST PERFECT FOOD

Jennifer Sartell

VOYAGEUR
PRESS

Inspiring | Educating | Creating | Entertaining

Brimming with creative inspiration, how-to projects, and useful information to enrich your everyday life, Quarto Knows is a favorite destination for those pursuing their interests and passions. Visit our site and dig deeper with our books into your area of interest: Quarto Creates, Quarto Cooks, Quarto Homes, Quarto Lives, Quarto Drives, Quarto Explores, Quarto Gifts, or Quarto Kids.

First published in 2017 by Voyageur Press an imprint of The Quarto Group, 401 Second Avenue North, Suite 310, Minneapolis, MN 55401 USA. Telephone: (612) 344-8100 Fax: (612) 344-8692
QuartoKnows.com

Voyageur Press titles are also available at discount for retail, wholesale, promotional, and bulk purchase. For details, contact the Special Sales Manager by email at specialsales@quarto.com or by mail at The Quarto Group, Attn: Special Sales Manager, 401 Second Avenue North, Suite 310, Minneapolis, MN 55401 USA.

10 9 8 7 6 5 4 3 2 1

ISBN: 978-0-7603-5222-9

Library of Congress Cataloging-in-Publication Data

Names: Sartell, Jennifer, author.
Title: Epic eggs : the poultry enthusiast's complete and essential guide to
   the most perfect food ever / by Jennifer Sartell.
Description: Minneapolis, Minnesota : Voyageur Press, 2017. | Includes index.
Identifiers: LCCN 2017006872 | ISBN 9780760352229 (pb)
Subjects: LCSH: Chickens. | Eggs.
Classification: LCC SF487 .S245 2017 | DDC 636.5—dc23
LC record available at https://lccn.loc.gov/2017006872

Acquiring Editor: Dennis Pernu
Project Manager: Jordan Wiklund
Art Director: Brad Springer
Cover Design: Faceout Studio, Spencer Fuller
Layout: Laura Shaw Design

Printed in China

CONTENTS

PREFACE

The Morning Egg

RAISING YOUR OWN POULTRY can be a rewarding experience. What I truly appreciate about it is the day-to-day rhythm and the little moments spent with our flock, observing their nuances, the sounds they make, and how they shape the spirit of our farm.

As the sun climbs over the horizon, it filters into our bedroom window. I hear our rooster's faint crow. He's locked in the coop, as I'm still in bed and haven't made my way down to open the door. A sleepy smile spreads across my face as I picture him, all business, strutting and strapping, taking very seriously his job of alerting the world that a new day has come, as if he were the one personally drawing up the daylight.

The hens will be waking now, too, and fussing around the feeders, enjoying their breakfast, their fat bottoms up in the air as they scratch and pick at food particles. They will be making their way to the nest boxes, eyeing each one from below with a tilt of the head as they decide which cubby will be used today.

It's springtime, so the whirr of incubators and the peeping of chicks in the brooder fill the empty silence of the farmhouse morning. These sounds mingle with the burping of the coffeepot and the hum and crackle of the last few fires of the year in the wood-burning furnace. They are the song of home.

Chicken keepers are doers; they actively participate in life. They enjoy caring for an animal and benefiting from that relationship in the form of a tangible, useful product. They work for their chickens and are paid in eggs. It's a respectful relationship balanced in each species nurturing the needs of the other. A barter—we feed the chicken and the chicken feeds us.

So I pull on my boots and my lightweight coat, grab the chicken-shaped wire egg basket, and head to the coops to take care of everyone for the day. Upon opening the door, a cacophony of clucks and throaty sounds greets me. The ducks in the adjoining coop hear me with the chickens and quack in a way that almost sounds like they're laughing. The geese scream their highest-pitched honks, which send our tom turkey into a shock gobble. The guineas, maybe the loudest of all, echo with a continuous banter of short, staccato calls. It's the most enthusiastic greeting I'll have all day!

EPIC EGGS

I top off the water and feed and check out the nest boxes. Three fresh eggs sit like treasures in a chest: one a lovely peach color, one a light green, and one a rich chocolate brown, still warm from the hens, and even a damp spot from the laying process on the peach egg. This is truly the farm-fresh experience. There will be more as the day goes on. I'll check again in the evening, when I bring the dinner scraps out and put everyone to bed for the night. These three eggs will be my breakfast.

In the other coop, I find two oblong duck eggs that will be perfect for baking, a guinea egg, and a turkey egg that I will add to the collection meant for the incubator.

With my basket of oval jewels, I head to the farmhouse for breakfast.

As an egg slides from the cracked shell into the oiled pan, the delicate white starts to sizzle and pop. I can hardly wait to dip my toast in the soft-cooked yolk, which sits pert, perky, and orange in the center of the egg. A farm-fresh yolk is more like a pocket of rich, buttery sauce, silken in texture and tasting truly of egg, a flavor that many store-bought eggs lack.

Farm-fresh eggs are delicious. Once you eat one, you'll have a hard time enjoying commercial eggs. You will find yourself saying things like, "These taste like rubber," or "Just look at this yolk—it's flat, pale, and runny." Those who don't raise poultry will think you're an egg snob. They may roll their eyes behind your back and secretly wish you would eat your breakfast in silence.

Which is fine. It's our job as poultry keepers to educate the world and to let people know what they're missing. Give them a dozen of your eggs and soon they'll be joining in the breakfast accusations.

Farm-fresh eggs are delicious. Once you eat one, you'll have a hard time enjoying commercial eggs.

INTRODUCTION

An Old New Fad

IT'S BEEN SAID that keeping backyard chickens is a fad. If so, it is a "fad" that is as old as written language.

Chickens were first domesticated around 7000 BCE in India and China, which means that we have shared a relationship with this animal for more than 9,000 years now. With that kind of shared history, it's hard to think of chicken-keeping as a fad.

For myself and many others today, keeping chickens is one aspect of a return to small-scale sustainability that is often described as a "back" to the land movement, or a "return" to the land—which itself is sometimes called a fad. But it's not as though that land we are returning to didn't exist all along. Now, as always, there is something inherently human about raising one's own food. There's a reason we are drawn to the

For many, keeping chickens is one aspect of a return to small-scale sustainability often described as "back to the land." Now, as always, there is something inherently human about raising one's own food.

beauty of gardens and the nostalgia of chickens dotting the pastoral landscape. Traditionally, these bucolic scenes are woven into our history, are a part of our poetry, literature, and art—the very things that define our humanity.

As a modern species, we've tried the faddish "new ways." We gave the microwave a shot; we tried out prepackaged, overprocessed, fast food; eaten meat pumped with antibiotics and growth hormones. We gave it a few decades, and many are deciding that the trial period is over, that this artificial lifestyle is not for us. We've heard the call: that quiet instinct that still whispers to our civilized selves. It is the voice of the generations who tended and cared for the food that they ate, from a time where food was respected, nurtured, and romanced.

This is not a fad at all, but simply a rediscovery of what has always been present. We are relearning a classic language as we renew lost crafts.

This rediscovery can be seen wherever we look. Some of us answer the call through a garden, by running our hands through the soil from which we came, planting a small seed, and watching it grow and present fruit to feed our

bodies. Gardening, in fact, is the doorway many of us pass through in the process of rediscovery, the action that creates the confidence to try other ventures like raising bees—or keeping chickens.

Preparing a meal with ingredients from your own backyard is incredibly satisfying, and that satisfaction is not accidental. It is a practice that coincides with our past. We are homesick for the traditions of our ancestors—the planting of seeds, the harvesting of produce, the simple, daily chores of scattering a handful of oats to a flock of cooing hens, the soft tactile pleasure of a chick covered in soft down, and the careful balancing act of retrieving eggs from a nesting box. These are the movements of our grandparents and their parents before them, the daily motion of sustainability that was the essence of life for generations past.

THE HERITAGE MOVEMENT

The terms *heritage* and *heirloom* are not new words, but recently they've been making their way into our language, literature, and daily life—very often in the context of a back-to-basics lifestyle. The terms can be seen on flower and vegetable seed packets, at farmers' markets and co-op groceries, and they are slowly filtering their way into corporate supermarkets. It is now part of the vocabulary of home poultry raising too. The terms *heritage* and *heirloom* apply to poultry and eggs in much the same way that they do for vegetables. And you can understand much about chickens and eggs by looking at recent trends in vegetables.

An heirloom vegetable is a variety that has not been hybridized—that is it's open pollinated. It is often a much older variety than the modern

The Jersey Giant is the largest chicken breed. It was developed as a dual-purpose bird, meaning it's a nice size for the table and it regularly lays large brown eggs.

cultivars. Many of these older varieties are gradually getting lost as scientific cross-breeding (hybridization) produces varieties with desired attributes: the new types may have very large fruit, or a pleasing shape, or smooth skin, or a long shelf life. But nothing is gained without giving something up, and modern hybrid fruits and vegetables have very often lost good attributes as selective breeding favored other qualities. As a result, modern hybrids may have lost the taste and depth of flavor that were found in the old heirlooms.

For example, your local supermarket likely carries only three or four types of tomato: large, red slicing tomatoes; Roma tomatoes for sauces; and perhaps one or two cherry types. The tomatoes are all uniform in size, shape, and color; they look beautiful lined up on the supermarket shelves, but they are often lacking in flavor. This limited range of tomato varieties is what consumers have settled for over the last few decades.

But when you extend the range into heirloom varieties, you will find that there are orange, yellow, purple, and green tomatoes, some with large bumps and craters, some shaped like light bulbs, and others with stripes. And each of these was originally bred for specific reasons—particular flavor, texture, sweetness, acidity, or structure. The most exciting revolution in produce today is that consumers are rediscovering the original heirloom varieties of tomatoes and other vegetables. At a good farmers' market or co-op grocer, these old varieties are now becoming popular again.

Exactly the same thing has happened in the livestock world, and especially in the chicken-and-egg world. Over time, the pressures of commercial breeding have gradually caused the disappearance of some species of cattle, pigs, and chickens in favor of those that produce large quantities of meat or milk or eggs. Now these heritage species are being rediscovered by consumers waking up to their advantages.

For some years now, the commercial chicken world has focused on just two breeds. For meat,

A complete meal mostly made from ingredients from our farm: scrambled, farm-fresh eggs cooked in goat butter with scallions from our herb patch; fried red-skinned potatoes from the garden; and homemade maple sausage seasoned with homegrown herbs and maple syrup from our sugar bush. Preparing a meal with ingredients from your own backyard is incredibly satisfying.

it is the Cornish Cross, a fast-growing cross of a Cornish chicken with a White Rock. For eggs, it is the Leghorn. The white eggs that you purchase at the supermarket most likely come from a Leghorn hen. While the Leghorn is still considered a heritage breed, it has all but monopolized the large-scale egg market. The rest of the breeds are left to the interest of backyard chicken keepers. It is these other breeds that make the study of eggs, and the raising of poultry, so interesting and so . . . epic.

There are, in fact, hundreds of breeds of chickens. Many have been named from the place where they were created, such as the Jersey Giant or the Rhode Island Red. Each breed was created with a specific goal in mind, whether it was for delicious meat, egg-laying ability, egg size, egg flavor, egg color, or broodiness—or

simply because they are beautiful. Although these species are not practical options for commercial farmers, for the backyard chicken keeper, they are exactly the point.

Raising chickens and collecting eggs have never been so interesting or exciting.

(opposite top) Consumers are rediscovering heritage breeds like the dual-purpose Rhode Island Red.

(opposite bottom) While the Leghorn, with its white eggs, is still considered a heritage breed, it has all but monopolized the commercial market.

WHY CHICKENS?

I find that keeping chickens is as much a statement
of lifestyle as it is a method for providing healthy food
protein. It is a philosophical, aesthetic endeavor
as much as it is a practical activity.

A basket of different-colored eggs can be as beautiful as a bouquet of flowers.

For many of us, a full basket of fresh, brightly colored eggs brings the same satisfaction that a philatelist might feel placing a rare find in his book of stamps. Even if you set aside the value of an egg as a food item, a basket of eggs is a beautiful thing. Like a bouquet of flowers, the colors and pretty, round shapes bring me joy.

For my own family, many of the chicken breeds we raise are kept for the variety of egg colors. I love the mix of colors when I collect eggs, and sometimes leave our egg basket out as a centerpiece on our front porch table. I also appreciate the aspect of self-sufficiency that a backyard flock provides. Raising your own food is a good way to gain control over the quality and

Many of the breeds we raise on my own farm are kept for the variety of egg colors. I sometimes leave our egg basket out as a centerpiece on our front porch table.

(opposite) The docile Australorp is a heritage breed—and a prolific egg layer.

variety of the items you eat. In keeping a flock of chickens, you decide what feed your chickens consume, how they are treated throughout their lives, and how the eggs are handled. If you want pasture-raised birds, or to raise them on organic feed, that's your decision. If you want to eliminate soy from their diet, you can. You can provide fresh vegetables, decide on rotation of pasture, determine which vaccinations and supplements they receive . . . all of which gives you control of what you put into your body.

Beyond the functional, practical, and aesthetic reasons for raising chickens, it should be mentioned that chickens are hilarious! In the evenings my husband and I love to sit in the yard and watch the chickens be chickens. It's more entertaining than any television program. Your flock will quickly form a small community with observable relationships and unique personalities, and you'll quickly establish a heartfelt connection to the little birds that coo and cluck around your yard.

And chickens make great pets. They can form emotional bonds with humans, and can be trained in many ways. They also provide an educational experience for all ages.

The Buff Orpington is my favorite breed of chicken. They are as gentle as teddy bears and make great chickens for children to be around.

The yolk of a free-range egg is usually bright orange and sits round and perky.

AND THEN . . . THE EGGS

Farm-raised eggs are delicious! I didn't realize what an egg could taste like until I started raising my own chickens. As a child and young adult raised on store-bought eggs, I was never a big egg eater. I considered them a somewhat bland, rubbery-like breakfast food that was greatly enhanced by butter and lots of salt and pepper, or scrambled with a good amount of cheese, veggies, or meat to make it taste like . . . something. Eggs for me were a sort of mild-flavored binder—a glue that held sausage, chopped onion, and some green peppers together in the shape of a western omelet.

Since raising my own chickens, though, I've come to realize what an egg is supposed to taste like. Eggs have a flavor! When cooked over-easy, the white has a delicate texture that almost melts in your mouth, and the yolk is rich and buttery, like a delicious sauce.

Fresh eggs look different also. The whites stay together when cracked in a pan, and the deep golden orange yolks sit high and rounded. A hard-boiled fresh egg will not have that grayish ring around the outside of the yolk. Farm-fresh eggs taste and look better because homegrown chickens are usually exposed to plenty of sunlight, fresh green grass, weeds, and protein-rich insects. All of these factors contribute to vitamin-rich, delicious eggs.

And of course, they are fresh! There are few things better than a warm egg taken from the nest box, brought directly into the house, and cooked to perfection. The integrity of the egg is lost as the egg ages, and refrigeration also diminishes the quality of an egg.

But the benefits of raising chickens go far beyond flavor. The number-one reason that I raise chickens is the satisfaction I get from a basket of eggs. Even more than the delicious taste, I'm motivated by the thrill of going out to the coop and collecting the amazing bounty that rests in our nest boxes.

Even more than the delicious taste of eggs, the thrill of going out to the coop and collecting the amazing
bounty is many a poultry keeper's number-one reason for keeping birds.

In the summer, we let our chickens free-range. They eat grass and bugs and get plenty of sunshine. As a result, summer eggs are abundant and delicious.

THE ART OF RAISING CHICKENS . . . AND EGGS

Did you know that just like tomatoes, eggs are seasonal?

In nature, a chicken lays the most eggs in the spring. This flow continues through summer and usually stops altogether in the fall, when the birds molt. Egg laying is directly related to the number of daylight hours, so when the molt is over, the chickens often settle in for a break during the winter months. Nature intended for a chicken's body to work this way. Eggs are a biological means for producing young, and a chicken lays eggs to continue her species, not to provide us with breakfast. So a chicken doesn't naturally produce offspring in the dead of winter.

During the winter months, a chicken's system can be tricked into thinking it's still summer by providing additional light. With an artificial light source, chickens will continue to lay throughout the winter. However, the eggs will not be the same as the eggs of summertime, just as strawberries purchased in February do not taste like fresh-picked strawberries in June.

In the summer, our chickens free-range. They eat grass and bugs and get plenty of sunshine, and as a result, summer eggs are abundant and delicious. The yolks are bright orange, rich, buttery, dense, and full of nutrients.

Geese will honk an alarm call when threatened. They will also chase off most medium-sized predators.

WHAT ABOUT OTHER POULTRY?

I've raised chickens, waterfowl, and other poultry for 21 years, with some breaks when I went to college. And I can attest to the fact that life is just better when you have a coop of poultry in the backyard. Chickens are by far the most popular backyard poultry, for several reasons.

- Chickens are readily available. Each spring, local feed stores get in a wide variety of chicks, which are very inexpensive and irresistibly cute! Chicks can also be ordered from hatcheries or from backyard breeders. Ducks, guineas, turkeys, quail, and geese may be a bit harder to locate, though as the self-sufficiency movement gains momentum, many of these birds are also becoming popular.

- Chickens are available in a wide variety of breeds. Raising chickens is much like keeping a collection. The breeds are so interesting and varied that it's easy to want one or two of each. Different species lay different-colored eggs, much the way that heirloom vegetable varieties can produce purple beans, yellow tomatoes, or blue corn.

- Chickens lay eggs consistently. Unlike some other breeds of poultry, chickens are reliable layers.

- Chickens are a manageable small size, easy to care for, and relatively quiet. A modest coop will easily hold three or four laying hens. A backyard chicken keeper can raise a small flock with very little initial investment. Larger breeds of poultry, such as turkeys, need lots of space and a large coop.

- Chickens are also easy to care for. Food, water, shelter, and a clean living space are the basics of chicken care. In many ways, chickens are easier and less expensive than caring for a dog or cat.

- We're familiar with chickens. People know chickens, and that makes them a comfortable animal to raise. We know what their eggs look and taste like, and that's appealing to many who are just starting out.

However, don't ignore other poultry species. As you gain experience, you may well want to consider other birds, because their eggs offer special merits.

DOES SHELL COLOR ACCOUNT FOR TASTE?

We raise chickens that lay green, olive, blue, and brown eggs. One of the most common questions I get when people see our eggs is, "Do the colored eggs taste different from white?"

No. Shell color has no effect on egg taste. The flavor of an egg is more attributed to the chicken's diet and lifestyle.

Khaki Campbell ducks are prolific egg layers.

Guineas are ravenous for the pests in your yard, gobbling up ticks, slugs, and even mosquitoes.

Our Black Spanish tom follows us around like a dog.

Quail

Quail lay tiny, delicious eggs that make any meal a culinary masterpiece. The birds are small and take up little room.

Ducks

Ducks are a manageable size, similar to chickens. They don't scratch like chickens do, so they work well when kept in raised-bed gardens. Some breeds, such as the Khaki Campbell, will lay as many eggs as a high-producing chicken.

Guineas

Guineas are excellent at pest control and are often kept to reduce tick populations.

Geese

Geese are wonderful guard animals. We've had a significant drop in predator attacks since we added geese to our flock.

Turkeys

Turkeys are not just for meat. They can make excellent pets. Our tom turkey has the personality of a dog: he follows us all over the yard, comes when he's called, and enjoys being petted. Turkeys are gentle giants, moving slowly with regal deliberation.

If having a pet turkey isn't reason enough to add one to your flock, consider raising a heritage breed that is in danger of extinction. You'll be surprised at how many people are looking to buy heritage hatching eggs or young poults.

GETTING STARTED WITH CHICKENS

Chickens are easy and inexpensive to care for. The largest investment in chicken keeping comes in the cost of the run and coop. Once this is built, daily upkeep is simple, and little is required to keep your chickens happy and healthy. But a little background information will help make chicken keeping a joy for both you and your flock.

Step 1: Check on the Legalities

The first step is to make sure that chickens are legal in your area. I've read countless, heart-breaking stories about people attempting to raise chickens in neighborhoods or cities where they are not allowed. Called before a city council or township board to make their case, the families are all too often forced to give up their flocks. It is much better to know what your community allows before beginning.

If your area doesn't allow chickens, I encourage you to petition your lawmakers or association members. Having control over what you eat is a right, not a privilege. Many times, educating your civic leaders is the best way to win your case. Three or four backyard hens are less noisy, less work, and create less odor than a backyard dog or two. Teach your community the benefits of keeping chickens. Many communities are becoming more open to the idea, which is a step in the right direction!

Step 2: Decide on Breeds

The next step is to research breeds. There are hundreds of breeds of chickens, each developed and domesticated for a different purpose. Some chickens are productive egg layers; others lay eggs with unique, interesting colors. Some were bred for meat, and others, like the Polish with its crested head, have beautiful plumage that is really the main reason for keeping them.

Another thing to consider is the size of the breed. If you live in an area where you can only keep a small coop, you may want to go with a smaller breed or even a Bantam—a miniature chicken that lays small eggs. Bantams require less space, less feed, and less upkeep. A large chicken, such as a Jersey Giant, may not do well in a small backyard coop.

Research which breeds do well in your climate. For the most part, chickens are hardy animals and will thrive in any weather, but some do better in harsh winters or in warmer climates. Chickens with rose combs or combs that are close to the body are less prone to frostbite and tend to do better in colder regions. Some breeds have heavier plumage, such as the Wyandotte, which protects them from the cold. Other chickens, such as the Turken or Naked Neck, do well in warm climates; the lack of plumage helps keep the animal cool in warm temperatures.

The last thing to consider is the general personality of the breed. You will always get exceptions to the rule, but breeds tend to have predictable personalities. In general, large chickens are more domesticated; they tend to be more docile, slower moving, and calmer. Good examples are Orpingtons, Cochins, and Brahmas. Smaller breeds, such as Bantams and Leghorns, tend to be flighty and wilder.

LOCATING HERITAGE BREEDS

Rare breeds are, well . . . rare. Raising a heritage breed is an admirable intention, but the birds are often difficult to find. Thanks to the Internet, the web of chicken breeders is growing and remote breeds are becoming more available. The Livestock Conservancy is a great place to start. They have information on breeds that need help as well as contact information for breeders who can offer chickens, chicks, or hatching eggs.

The Polish breed boasts a striking crest of feathers around its head.

Bantam breeds are smaller and require less space and feed. However, they're known to have uppity personalities.

This Blue Laced Red Wyandotte has a rose comb that sits close to her head, which makes it less susceptible to frostbite.

(above) This triangle-shaped coop and run is perfect for three chickens. The fine-mesh hardware cloth works much better at predator-proofing than chicken wire does.

(right) This coop holds our collection of breeds that lay an array of differently colored eggs: Leghorns (white eggs); Easter Eggers (green, pink, and blue); French Black Copper Marans (dark brown); Buff Orpingtons (light brown); and Olive Eggers (olive).

Step 3: Plan Your Coop

The nice thing about raising chickens is that it's almost as easy to raise three as it is to raise six or twelve or twenty. Once you have the coop, the feeders, and the waterers, it's just as easy to scoop out feed for ten chickens as it is to feed two. You are only limited by the size of your coop and run.

Chickens have very basic needs. They need a place to live—a place that's protected from the elements and predators where they can roost at night, lay eggs, and eat and drink. There are a variety of chicken coop designs to choose from. You may want a coop with a run, or perhaps you plan on free-ranging birds. Or you may want a chicken tractor that can be moved around the yard, providing a new, protected grazing area each time.

Chicken coops can be elaborate, beautiful examples of custom-built architecture, or they can be made be made from all kinds of recycled materials or structures: recycled doghouses, old playhouses, sheds—even out-of-commission cars or old furniture—can be converted into a chicken living space.

Whatever you decide to use, every chicken coop should have these elements:

- **Space** Chickens do well with at least 3 square feet per chicken in the coop, plus run space. Larger breeds may require more; smaller breeds, less. Chickens will always appreciate more space if allowed. Overcrowding can result in territorial issues, pecking, and, in extreme cases, cannibalism.

- **Shelter from the weather** Most chickens can tolerate cold temperatures. Our chickens do fine in Michigan, where winter days can drop well below 0°F. I never recommend

Chickens will instinctively return to their coop each night at dusk, which makes it easy to lock them away from nighttime predators.

heating a coop. Not only is it a fire hazard, but the artificial heat can also cause condensation buildup, which can lead to respiratory issues and infection. Instead, make sure that your coop is well insulated against drafts and wet weather. Also make sure that your coop has sufficient ventilation to allow heat to escape in the warmer months and for odors to dissipate.

- **Protection from predators** For the most part, we free-range our chickens during the day and lock them up at night. We occasionally lose a bird to a daytime predator such as a hawk or fox, but this has become very rare since we added geese as guardian animals. For us, the joy and health that our chickens enjoy from free-ranging is worth the risk of an occasional predator attack. But this decision is one that each chicken keeper has to make.

    Chickens have a natural homing instinct to return to their coop each night at dusk. Ducks, geese, and turkeys might need extra training, but they can be taught to do the same. The flock should be locked away safe in the coop throughout the night to protect against nocturnal predators, such as raccoons, owls, and opossums. If you have a run, the run should be covered and reinforced along the ground to protect against digging predators, such as foxes and coyotes, and should have an overhead cover to protect against hawks, eagles, and climbing predators such as raccoons.

- **Nest boxes and roosts** One nest box should be enough to serve four to six hens. If you allow too many nest boxes, the chickens will sleep in them and defecate in the boxes all night, resulting in dirty eggs.

    Chickens have the instinct to roost when they sleep. In the wild, they would use tree branches to protect themselves from ground predators. If you provide roosts, they will happily perch there each night.

(above) Chickens make egg collection easy for their keeper. They will lay their eggs in nest boxes, making gathering eggs as simple as bringing in the mail.

(left) One nest box will typically be sufficient to serve four to six hens.

- **Bedding** To keep the coop fresh and easy to clean, some sort of bedding should be used. We prefer kiln-dried pine flakes. The coop should be cleaned regularly to keep odors, diseases, and parasites from becoming a problem.

Step 4: Establish a Food and Watering Plan

As chickens become more and more popular, more and more chicken feeds are coming on the market. Some are designed for young chicks, some for young pullets and cockerels, some for meat birds, and some for laying hens. There are also organic feeds, feeds designed to encourage feather growth, feeds that add omega nutrients to eggs, show bird feed . . . the list goes on. You also have the choice of feeding mash, crumbles, pellets, or whole-grain feeds. Some chicken keepers decide to forgo feed altogether, relying instead on rotational pasture feeding to raise their flock.

Whatever you decide, chickens need certain nutrients at different ages. The easiest way to provide those nutrients is to feed with a supplemental, commercial food. Protein is a main consideration when choosing feed.

Whatever you decide—mash, crumbles, pellets, organic—your birds will need specific nutrients at different ages.

- Newborn chicks to 8 weeks: starter feed, 20 percent protein

- Young birds 9 to 20 weeks: grower feed, 18 percent protein

- Laying hens 21 weeks and older: layer feed, 16 percent protein plus calcium

Non-free-ranging birds will also need free-choice grit to help digest their food. "Free-choice" simply means the grit is left out for the birds to ingest as they need it. Laying hens also benefit from a free-choice calcium supplement to ensure healthy eggshells.

In addition to a quality feed, fresh, clean water is very important to a healthy flock. Chickens should have access to water at all times. There are many waterer designs on the market: vacuum-type systems, nipple waterers, and, perhaps the simplest, a shallow dish. In the winter, a heating device should be provided to keep water from freezing.

There are many ways to spoil chickens by adding supplements, treats, and elaborate housing, feeding, and watering systems. For the most part, chickens need very little to keep them happy and laying.

A trio of adolescent Brahma pullets enjoy a grower feed with 18 percent protein.

THE HUMBLE EGG

◈

Until I started raising chickens, I didn't appreciate eggs. I didn't give them the credit that they deserve. They were an everyday item that was added to my grocery list, something to eat with bacon or add to a cake recipe. I didn't see them as a capsule of potential, something capable of perpetuating a species and bringing forth a new generation. The egg is humble, concealing its wonder in a quiet stillness, until one day it bursts open with life.

Different chicken breeds are known to lay a variety of egg colors, ranging from white to light peach to dark chocolate brown and even blue, green, and olive.

To the naked eye, the egg is a simple enough object—a seemingly inanimate ovoid, plain in texture, symmetrical, and common. But what an egg is capable of creating is nothing short of a miracle.

When a mammal grows an offspring, the mother's body undergoes all sorts of complicated chemical changes and developments. She feeds the growing embryo with nourishment from her own intake, while her hormonal systems engage and cater to each stage of delicate fetal growth. Her body is completely overhauled to sustain the life inside her.

The egg, however, takes all those complicated systems, wraps them neatly in an oval shell, and leaves nature to take care of itself.

And it does . . . quite well, in fact.

With the exception of warmth and humidity, the egg contains everything a growing embryo needs to become a chick. It is, in many ways, similar to a seed. The egg can be kept fertile, yet dormant, for a period of time, but when warmth and moisture are applied it begins to germinate.

Even the anatomy of an egg seems simple, consisting of three uncomplicated components that hardly seem capable of growing a living being: the shell, the white, and the yolk.

But an egg is extremely efficient with its parts. The shell, for example, is not just a solid covering but also features thousands of microscopic pores and layers of film that regulate the amount of oxygen and moisture that enter while protecting the chick from bacteria. In fact, every aspect of an egg is perfectly designed to work in a nurturing relationship with the chick.

Beyond the shell, white, and yolk are more specific parts of an egg that encourage life.

EGG ANATOMY

To those of us who love to eat eggs, the yolk and white are delicious accompaniments to buttered toast. But an egg is capable of much more. An egg has ten main components that make up its functioning anatomy.

Shape The egg's oval shape allows the hen to turn the chicks, but is not so round that it has a tendency to roll away. Mathematically, the egg is a strong shape. Even while its shape provides strength to protect the growing life inside, the egg's shell is delicate enough so that an unborn baby chick can break through. The egg is a perfect blend of durability and fragility.

Bloom The bloom is an invisible layer with which the hen coats the shell as she lays the egg. The bloom helps prevent bacteria from passing through the microscopic pores of the eggshell.

Shell The eggshell is made of calcium carbonate, which gives it a matte, chalky texture. The shell has thousands of microscopic pores that allow air and humidity to flow in and out.

The outer and inner membranes These layers just inside the shell wall are made mostly of keratin, a protein that also gives structure to human hair and is quite strong. If there isn't sufficient moisture when a chick hatches, these layers can dry out and become rubbery, making it difficult or impossible for the chick to break through.

Air cell The air cell forms between the inner and outer membranes. When a chicken lays an egg, it is warm, the temperature of her body. As the egg cools, the inner membrane contracts away from the outer membrane and a layer of air is formed, called the air cell, that insulates the egg. The bottom (wide end) of the egg usually has a large pocket of air.

Chalazae cords   If you've ever cracked open a fresh egg, most of the white is transparent. Often you will see a more opaque strand of twisted substance stemming from the yolk. These are the chalazae, cords of twisted protein that hold the yolk in the center of the egg. The chalazae are twisted in opposite directions to provide strength and stability to the yolk. As an egg ages, these cords break down, so more obvious chalazae indicate a fresher egg.

Outer and inner albumen   When you crack an egg, you might notice an oval of more substantial white that seems to hold its shape, as well as a runnier area of the white that spreads out in the pan. This runny white is the outer albumen; the thicker oval of white is the inner albumen. The albumen acts like a layer of bubble wrap, creating a protective barrier around the yolk and delicate embryo. The thicker viscosity of the inner albumen lends extra stability.

Vitelline membrane   If you've ever tried to separate yolks from whites to make meringue or custard, you know how important it is not to break the thin vitelline membrane, which acts as a clear casing around the yolk and keeps the contents intact.

Germinal disk   For chicken breeders, it is important to know whether your rooster is fertile and successfully inseminating your hens. One way to check this is to look for the germinal disk. This is the spot where the sperm entered the egg, and it appears as a white circle on the yolk. The chick will begin to grow from this point. Blood vessels will form and spread out into the yolk, gathering nutrition to feed the embryo.

Yolk   This golden pocket of nutrition contains all the fat, protein, and vitamins a chick needs to reach hatching age.

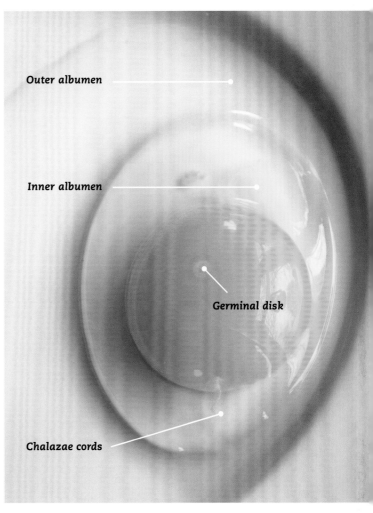

The germinal disk is a small white circle on the outside of the yolk that indicates where sperm has entered the egg. The inner albumen is a thick layer of white, while the outer albumen is of a thinner consistency. The chalazae cords appear as white strands stemming from each side of the yolk.

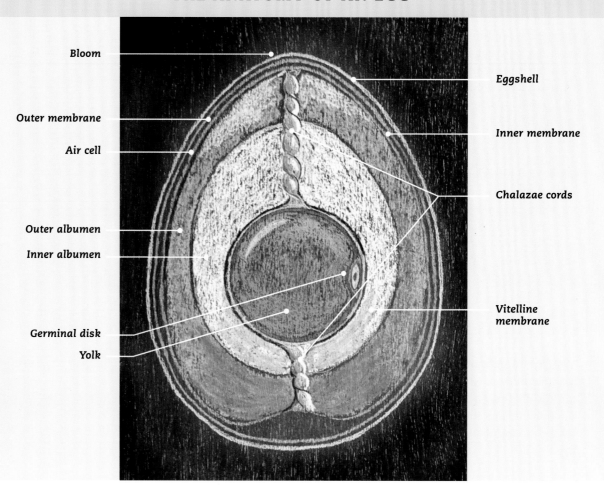

Bloom

Eggshell

Outer membrane

Inner membrane

Air cell

Chalazae cords

Outer albumen

Inner albumen

Vitelline membrane

Germinal disk

Yolk

THE MAKING OF AN EGG IN THE OVIDUCT

So how is this amazing object created? Let's take a look at the formation of the egg inside the hen.

The oviduct is the complete reproductive tract of a hen. It is around 31 inches long and includes the infundibulum, magnum, isthmus, uterus, and vagina. The oviduct acts like an assembly line: the yolk moves along and at each station it adds another layer, ending with the shell.

1. Ovary The ovary appears as the overall cluster of ova. Only the left ovary on a hen is functional.

2. Ova The ova appear as tiny yolks. Each one is encased in a sack and they are clustered together like grapes. As the ovary releases ova, they grow to the size of full yolks. A pullet chick is born with around 4,000 ova.

3. Infundibulum (funnel) The sack that keeps the ova connected to the ovary bursts, releasing the yolk into this funnel. If the hen has mated with a rooster, this is also where the sperm will enter the yolk.

HEN OVIDUCT ANATOMY

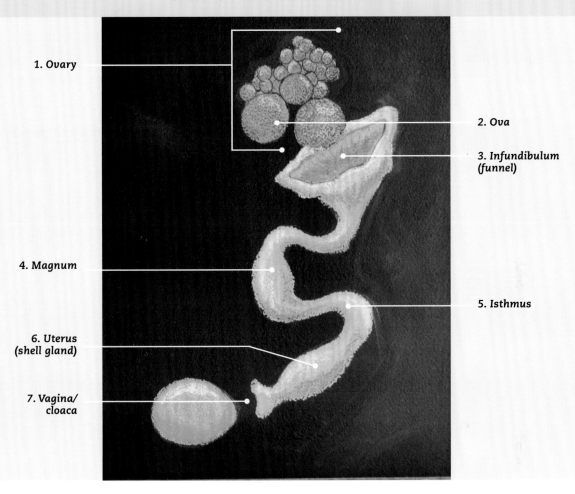

1. Ovary
2. Ova
3. Infundibulum (funnel)
4. Magnum
5. Isthmus
6. Uterus (shell gland)
7. Vagina/ cloaca

4. Magnum   Here, the yolk is surrounded by the albumen, or egg white.

5. Isthmus   The inner and outer membranes are added here.

6. Uterus (shell gland)   This is where all the color happens. The shell gland distributes the calcium carbonate that will harden into the shell once the egg is laid. This is also where the egg's pigment is created and distributed onto the egg.

7. Vagina (cloaca)   The egg enters the outside world through the vagina.

A chicken takes about 24 hours to lay an egg. The longest part of this process is the formation of the shell, which takes around 21 hours.

Although a hen is born with about 4,000 ova (or potential eggs), she will not lay nearly this many in her lifetime. In fact, the average hen lays 500 to 550 eggs in her lifetime. Chickens usually don't lay many eggs in their first year of life. You may get a few eggs before winter, as the pullet reaches maturity. After the first year, she will lay very steadily for about three years before production slowly tapers off.

SILKIE: THE MOTHER OF ALL CHICKENS

In addition to being beautiful birds, Silkie hens are serious brooders that make good adoptive mothers.

With their small stature, fluffy downlike plumage, bright-blue earlobes, and black skin, Silkies are often viewed as a "fancy" chicken breed. Serious producers might view them more as pets than as useful egg layers or meat birds.

But Silkies have unique characteristics that can be useful in the farmyard setting. They make excellent mothers! The Silkie's instinct to go broody is continuous throughout the warmer months. Silkie owners know that these persistent mothers will not only sit on their own eggs but on those of other hens as well. They are known to steal eggs from other chickens and even from other species.

I once had a Silkie that would steal turkey eggs. The turkey hen would get quite upset with the little chicken (understandably), but the Silkie stood her ground, fluffing up her feathers and squawking until the turkey backed down. Then the Silkie continued her mission, scooting the large turkey egg across the coop floor and adding it to her clutch.

Over the course of one week, she successfully stole three turkey eggs. Our fluffy little chicken sitting on big turkey eggs was quite the sight. She managed to hatch out a poult that year and raised it with the other chicks.

If you're not interested in setting up an incubator but want to renew your flock with young birds, consider adding a couple of Silkies. If you have specific eggs you'd like incubated, you can give them to a Silkie to hatch and raise to adulthood. This handy little chicken can save you from the responsibility of tending to an incubator and brooding young chicks.

EGG SHAPE AND SIZE

When you look at a basket of eggs, they're usually all roughly the same shape: oval, with one end wider than the other. But if you compare one egg to another, you might see that some eggs are fat and round, while others are thinner with a pointed top. (Incidentally, some claim that the shape of an egg is a sure way to tell the sex of the chick inside—pointed eggs are said to be roosters, while rounded eggs are supposed to become hens.)

As with shape, eggs also exhibit varying sizes. Sometimes, a newly laying hen lays smaller eggs her first spring after hatching. Her eggs should grow in size and weight as she matures.

Also, hens might lay smaller eggs during the winter and into spring or after a hard molt.

Eggs can vary widely in size and shape—sometimes from the same hen. This photo provides a good comparison between a normal-size egg from a seasoned laying hen and a smaller egg from a newly laying hen.

Some claim that a pointy egg like that on the left is indicative of a rooster inside, while hens hatch from round eggs. This has never been proven.

PLANTING THE CHICKEN SEED: INCUBATION REQUIREMENTS

A pumpkin seed needs warm soil and moisture to germinate. It also requires its parent flowers to "mate," or pollinate, thus producing the fruit that hold fertile seeds. Once a pumpkin seed is formed inside the fertile pumpkin, it holds everything it needs to germinate into a young seedling.

The hen-to-egg-to-chick process is analogous. The hen is the pumpkin, the egg is the seed, and the chick is the sprouted seedling. Once the seedling germinates, it spreads roots and takes things from its environment to grow, just as a hatched chick eats and drinks beyond the egg.

A hen, however, does not need a rooster to lay eggs. Her body continues to produce unfertilized eggs with or without a rooster present, though her unfertilized eggs will never hatch into chicks.

Egg incubation requires the egg to be kept at a consistent temperature of around 99.5°F and a humidity rate of around 50 percent. With these two conditions in place, the fertilized egg will begin cell division, resulting in the growth of an embryo. There are two ways to incubate an egg: with a broody hen and with an artificial incubator.

The Broody Hen

In nature, incubation starts with a broody hen. When a hen is broody, she is overcome by the instinct to hatch out young. She will find a good place to begin her clutch, or group of eggs. She might arrange the nesting material and move things about to her liking. Each day she will lay her egg in the nest until she has a good number (I've seen anywhere from five to nine eggs). If her clutch is too large, she can have a hard time keeping all the eggs at the right temperature, as only so many eggs will fit under her at once. This often results in underdeveloped embryos that don't live long enough to hatch. Once the clutch is laid, the hen will go broody and begin her 21-day sit.

A broody hen is overcome by the instinct to hatch out young. She will arrange nesting materials to her liking and even sit on eggs from other hens—or other species!

Her body also undergoes physical changes. She begins running an elevated temperature to help provide warmth to the eggs. She also plucks feathers from her breast to ensure skin contact with the eggshells. This is called a broody patch. During this 3-week span, the hen rotates the eggs several times a day to ensure proper development. And she stops laying eggs while she sits.

Her digestive system also changes while she is broody. During the 21-day gestation period, she leaves the nest only once a day to eat, drink, and eliminate waste. The droppings from a broody hen are huge. She becomes very aggressive and defensive of her eggs, puffing up her body to make herself look as large and intimidating as possible.

Because she sits on all the eggs at the same time, each egg in the hen's clutch should hatch around the same time. This timing is important because once the chicks hatch and dry out, they begin to use their legs. If chicks were to hatch on successive days, the hen would have a hard time protecting the hatched chicks while staying put on the unhatched eggs.

Once the chicks have hatched, the hen usually consumes the eggshells. This provides calcium and removes evidence for predatory animals. You can feed your mother hen the same protein-rich food that you feed the chicks. Her body will need extra nutrients and recuperation after the long sit. Her wattle may be pale for a while and her feathers might be rough. She will probably enjoy a nice dust bath after her chicks are old enough to follow her outside.

Some breeds are more likely to become broody. Silkies, for example, make wonderful mothers (see page 42). These hens need very little encouragement to start a clutch. I've even seen them hatch out other species like ducks and turkeys.

The Incubator

While I personally prefer it when a hen takes over the job of incubating eggs, it is not always possible. With many breeds, the instinct to become a mother has been bred out and often the only way to perpetuate the breed is to incubate the eggs artificially. Incubators also give the chicken keeper control over which eggs are selected to hatch.

My suggestion when purchasing an incubator is to get the best-quality incubator you can afford. I use an incubator that holds seven chicken eggs. I've also hatched turkeys and ducks out of these incubators.

There are three main considerations when artificially incubating fertile eggs:

1. Temperature control   Keeping the eggs at a consistent temperature of around 99.5°F (37.5°C) is very important to the development of the growing chick. This temperature can range a couple of degrees warmer or cooler for short periods of time.

When setting up the incubator, make sure it is clean from the previous hatch. Any nonelectric parts can be washed with soap and water, and more delicate parts can be disinfected with a cotton swab and rubbing alcohol (where appropriate). Thorough cleaning ensures that bacteria

Not all incubators have a turning mechanism. When hand-turning eggs, use a pencil to draw an X on one side of the egg and an O on the other side. Each time the egg needs turning, flip it 180 degrees, exposing the O or the X. Use a checklist to make sure the eggs are turned properly, and set your phone timer to remind you.

I prefer hens do the incubating whenever possible, but when that's not an option, an incubator like this does a good job of ensuring proper temperature and humidity for the eggs.

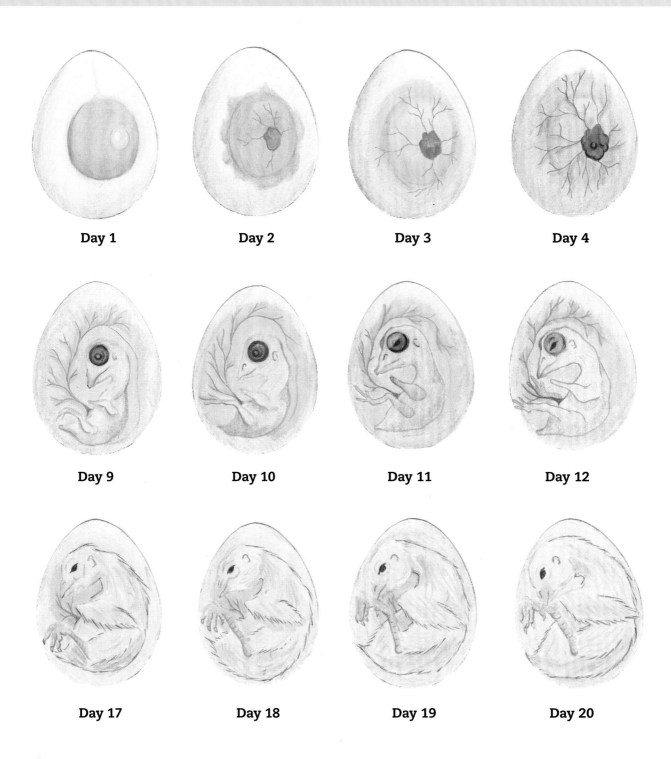

Day 1   Day 2   Day 3   Day 4

Day 9   Day 10   Day 11   Day 12

Day 17   Day 18   Day 19   Day 20

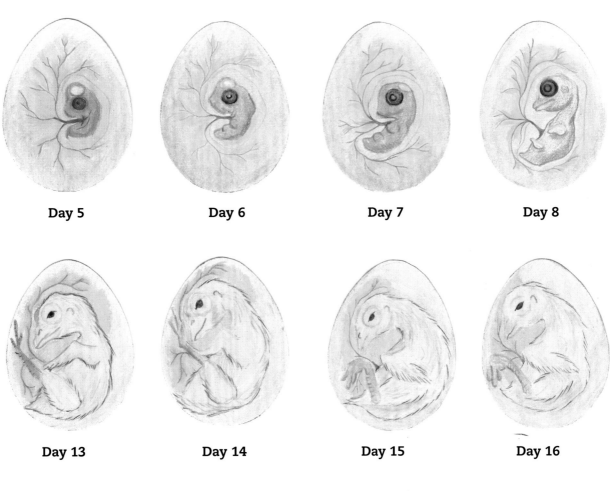

Day 5 **Day 6** **Day 7** **Day 8**

Day 13 **Day 14** **Day 15** **Day 16**

A chick's gestational period is 21 days. With such a short incubation time, development is quite rapid with significant changes taking place on a daily basis. When candling the egg, vein formation can be observed on day 4, and a heartbeat is visible by day 6. Feathers begin growing on day 10, and the wattles and comb are developing by day 13. Eventually, the chick will fill the entire egg and then begin the hatching process.

The Humble Egg   47

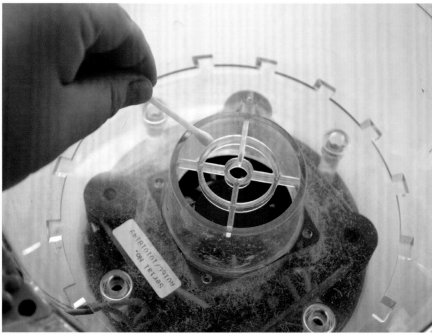

Make sure your incubator is clean from the previous hatch. Any nonelectric parts can be washed with mild soap and water. Clean more delicate parts with a cotton swab and rubbing alcohol.

growth from past hatches won't affect your hatch. Once clean, the incubator should run for at least a few hours to regulate the temperature.

Set up the incubator in a relatively temperature-controlled room. Don't place it by a door that's opened often, letting in cold breezes, or by a heater vent. Also be prepared to have a backup plan if you live in an area that loses power often.

2. Humidity   Humidity is very important to unborn chicks, especially in the last few days of development. It's crucial that the inner and outer membranes of the egg stay moist, or the strong band of keratin can become tough and rubbery. It will actually shrink and suffocate the chick, making it impossible for the chick to break through the eggshell. Humidity should be kept at around 50 percent on days 1 through 18, and then raised to 70 percent on days 19 through 21.

If your incubator does not give a digital humidity reading, you can purchase a hygrometer to help ensure proper humidity levels.

Do not open your incubator often. This not only changes the temperature but also allows built-up humidity to escape.

3. Egg rotation   Eggs must be rotated several times a day so that the growing chick does not adhere to the wall of the egg. Many incubators come with a rotational device that does this automatically. You can program it according to the circumference of the egg and how often you want the eggs rotated. We usually set it for every 45 minutes to an hour. Our incubator uses a rolling system where one disk rotates in a circle with small cutouts. Some incubators have separate rotation systems that use more of a leaning motion. The egg tray is rocked back and forth at regular intervals.

HOW TO CANDLE AN EGG

The egg is an amazing, natural phenomenon. Candling is a nonintrusive method that allows you to see the shadow of the chick's heartbeat and, if done daily, the amazing progress of a growing chick. To candle an egg, you need only a concentrated light source like a flashlight, a toilet paper roll, and a darkened room.

First, cut the cardboard paper roll in half to shorten its length. Next, place the egg at one end of the tube and shine the flashlight through the other end.

If you've had eggs in your incubator or under a broody hen, you should be able to see the formation of veins around day 4 or 5. After that, you should be able to observe the chick's growth and movements throughout the 21-day incubation period. If no signs of life or progression are present after day 7 or 8, remove the egg from the incubator. A dead or infertile egg will go rancid in the warm, moist environment and can even explode.

As you come to the end of the incubation, the chick will be large and taking up most of the space in the egg. If you candle at this time, you won't be able to see much of anything but darkness. But that's a good sign that your chick is large and ready to enter the world.

When candling, be sure not to open the incubator too often (once a day should be fine), and replace the lid between checking eggs. This keeps the moisture and temperature levels consistent. Also, make sure you don't keep eggs out of the incubator for more than a couple of minutes while candling.

Some companies sell devices that allow you to candle an egg without moving into a dark room. The Brinsea OvaScope, for example, uses a light and a mirror to reflect the inside of the egg back to the viewer. It also has a device to rotate the egg. This helps make egg viewing a bit safer because you're more likely to drop an egg balanced on a toilet paper roll in a dark room.

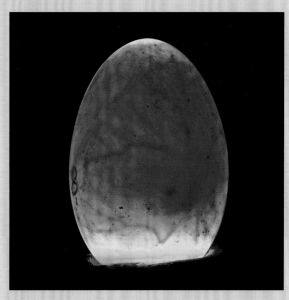

(above) A developing chick's vein formation is evident inside this candled egg.

(left) Devices such as this OvaScope make candling easier and more foolproof than "classic" methods, such as a flashlight and toilet paper roll.

Humidity control is crucial to unborn chicks, especially in the last few days of development. Resist the urge to open your incubator.

Yet other incubators don't provide a turning device, so you must set up a schedule for hand-turning the eggs. Eggs should be turned as often as your schedule allows, or a minimum of four times daily. I've found that the best way to set up a routine and keep track of turning eggs is to use a pencil to draw an **X** on one side of the egg and an **O** on the other. Each time the egg needs to be turned, it should be flipped 180 degrees, exposing the **O** or the **X**. You can also set up a table and checklist system to make sure the eggs are turned properly. Set your phone timer to go off at each interval.

In general, it's also a good idea to number your eggs. This way, you can keep track of which you've rotated or checked for development.

Two days before hatch, the rotation should stop. We remove the rotating disk in our incubator and replace it with a paper towel. This gives the hatching chicks a nonslippery standing area, which is important to the proper development of the legs. It also helps absorb hatching fluid and makes cleanup a bit easier.

The pip is the first crack in the eggshell made by the chick using the egg tooth on the end of its beak.

HATCHING

Day 21 has arrived! Waiting for eggs to hatch can seem like eternity. It's such an exciting time! On day 21 you should start to see some wiggling. The eggs might rock back and forth slightly, and you may also hear peeping through the shell. One thing to understand is that hatching can take several hours. It doesn't happen like in cartoons where the chick busts through the eggshell in one movement.

Hatching starts with the "pip," the first crack in the egg, usually a diamond-shaped hole. If you look closely, you will see the chick's beak. The chick uses its egg tooth, a sharp point on the tip of the beak, to make the initial pip. The chick may stay like this for an hour or two. You will hear it peeping through the pip hole, and fussing with the corners.

As the hatch continues, the chick will extend the crack from the pip hole around the circumference of the egg, bisecting the eggshell. Slowly, the crack grows until the shell is connected only by a hinge of shell membrane.

The chick uses force to push against the opposite ends of the eggshell, prying them apart

A newly hatched chick waits in the incubator for its feathers to dry or "fluff out."

and opening the crack. This is exhausting work. It may lash out with bursts of energy, pushing and wiggling, and then collapse with exhaustion. Being born is hard work! Resist the urge to assist a chick. This is a delicate process that the chick needs to undergo. Plus, the chick is still connected to the eggshell via veins. If you pry the egg apart you could kill the chick.

Once the chick has hatched, leave it in the incubator until it fluffs out (its feathers have dried completely). If the humidity level is too high from the added moisture of the hatching eggs for the chicks to dry out, wait until all have hatched before opening the incubator. That way, you don't release the much-needed humidity for the unhatched chicks. Then the chicks can be placed under a heat lamp set at around 100°F (38°C) to finish drying out.

(top, left) This photo of a Black Spanish turkey mid-hatch provides a good view of the inner egg's membrane and how the egg is cracked along its circumference, or equator.

(bottom, left) Hatching is exhausting work. A baby bird may lash out with bursts of energy, pushing and wiggling before collapsing in exhaustion.

BASIC BROODER SETUP

Chicks are among the easiest animals to care for. They need very little to thrive and the equipment is relatively inexpensive. The area where you raise chicks is called a brooder. A basic brooder setup includes:

Vessel   You need an enclosed container that offers plenty of ventilation but will also block out breezes and keep the chicks safe. A large plastic storage bin or even a heavy-duty cardboard box works well. My favorite brooders are large-animal water troughs. They are durable and can be used over and over for years. They're also waterproof, which is great for raising ducklings and goslings, which like to throw water everywhere.

Be mindful that chicks generate a lot of dust, mostly from the scratching that they do, which kicks up powdered feed. I recommend keeping your brooder away from the kitchen or rooms with food. Choose a room that can easily be dusted or vacuumed often. A spare bathroom works well.

If you live in a climate where the spring weather is mild, the brooder can be kept in a draft-free, dry outbuilding, barn, garage, or possibly even the coop in which the grown chickens will eventually live. Make sure the brooder is safe from predators, curious pets, and any adult chickens that you may already have. Keep a close watch on the thermometer to make sure the temperature doesn't fluctuate too much as the weather changes.

A Buff Orpington chick. Once your chicks arrive they have a specific, but relatively inexpensive, set of requirements.

Cardboard boxes and feed tanks work well as brooders. Keep your chicks safe and warm indoors, away from adult chickens, until they are old enough to be slowly integrated into your flock.

Waterer   I recommend a waterer designed specifically for baby chicks. The typical waterer is a vacuum system with a shallow tray designed so that chicks won't drown in it. You'll notice that many feeders and waterers have a red bottom. Because chicks are drawn to red, this helps them find their water easily. I give room-temperature water for the first week or so. The waterer should be cleaned with soap and water or a dash of white vinegar every other day to prevent odor and bacterial growth.

Feeder   There are several designs on the market. Some are inverted, similar to the vacuum waterer, and some are longer trays with holes cut out. Both are designed to help eliminate feed waste. Chicks love to scratch in their food and spread it out everywhere. To help prevent this further, elevate the feeders as the chicks grow so that only their heads can reach the feed.

Feed   Chicks can live without food or water for 48 hours after hatching because they are still being sustained by the nutrients in the yolk. So don't feel bad if they aren't eating or drinking right after they fluff out. When they *do* begin to eat, chicks need feed with a high protein content. This helps them develop properly. Buy a starter feed designed for chicks with 20 percent protein content.

Many feeds come medicated to prevent diseases like coccidiosis, a parasitic protozoa that affects the digestive system. However, if your birds have been vaccinated at the hatchery, medicated feed should not be given, as it will counteract the vaccine.

Chicks are born with a good amount of instinct, but some things they must learn from their mother. In the wild, the hen will find a morsel of food, pick it up with her beak, and drop it. By repeating this, she is demonstrating to her young where the food is and how to eat it. Because brooder-raised chicks don't have a mother hen to show them the ways of the world, you must step in and play mother hen in some instances. One of these is beak dipping.

To show chicks where their food and water is, take the chick in your hand and dip the tip of its beak in the water tray. Be careful not to go too deep. The chick will often throw its head back and drink down any liquid it may have scooped up. I then dip their beaks in their crumbled feed. Do this with every chick each day until you witness them eating and drinking on their own, which usually takes only a day or two.

Chick grit   Chickens don't have teeth to grind their food into small bits. Instead, they have a crop, a pocket beneath the esophagus in the digestive tract. Chickens consume small rocks and bits of gravel, which they store in the crop to help grind food into smaller, more digestible pieces.

Because your chicks are being raised in a brooder with manufactured bedding, they do not have access to dirt, sand, or small pieces of gravel. Therefore, you must provide them with grit. Chick grit can be purchased at your local feed store. Make sure it's designed specifically for chicks because the pieces are smaller. Grit can be lightly sprinkled daily as a top dressing on the feed.

Bedding   My favorite bedding material for the brooder is kiln-dried pine flakes found at most farm supply stores. The flakes are absorbent and give chicks a steady walking surface. The pine oils also lend a pleasing scent that masks odors. Finally, it's lightweight and easy to scoop away once soiled. Your brooder box should be cleaned at least once a week—all bedding removed and replaced with clean bedding. And avoid cedar chips around chickens. The cedar oil can produce eye and lung problems.

Heat lamp   Chicks need to be kept warm. Mother hen would still be sitting on them and keeping them toasty at this young age, but because your chicks are in a brooder, you again have to simulate mother hen, and this means using a heat lamp.

The most common and least expensive heat lamp is usually the large tin funnel type with a clamp or a place to hang the lamp. Be sure to secure the lamp well, as misplaced heat lamps have been known to cause fires.

Before the chicks are born, set up your brooder with the heat lamp. Use a thermometer to gauge the temperature. Adjust the heat lamp higher or lower to allow for a steady 100°F (38°C) in one area. The brooder should be large enough so that there is also a cooler area away from the heat lamp where chicks can move if they get too warm. Their water should also be kept in this cooler area.

Chicks and guinea keets in the brooder box. The trays on both the waterer and the feeder are red, a color to which chicks are drawn. In the brooder, I recommend waterers designed specifically for baby chicks.

For the first week, chicks should be kept at around 100°F (38°C). Each week, the temperature can be lowered 5°F.

You might notice that there are two different bulb choices for your heat lamp: red and white. Many people use red bulbs in their brooder as an extra precaution against overzealous picking. Chicks are programmed to establish a pecking order at a very young age and will practice pecking their brooder mates to communicate where each belongs socially in the flock. Sometimes this pecking can cause a small injury. Chicks are attracted to the color red, so if one of the chicks develops a small cut, the red color of the scab will be a magnet to the other chicks. They will continue to peck at the scab and, if left unnoticed, they can sometimes kill a chick. The red bulb turns the whole brooder landscape red, making it difficult for the chicks to pick out one particular red object. We've used both red and white bulbs over the years and have rarely had a problem with pecking. But if you want to be extra safe, choose red.

For the first week, chicks should be kept at around 100°F (38°C). Each week, the temperature can be lowered 5°F by raising the heat lamp an inch or two. Continue to decrease the temperature until the average outside temperature matches the temperature in your brooder, or until the chicks are completely feathered out (this stage happens at different times for each breed). It is now safe to place them outside, provided the outside temperatures are suitably warm.

Thermometer Always have a thermometer situated in the warmest place in the brooder so you can check the temperature. Another gauge for checking brooder temperature is chick behavior. When the right temperature is reached, chicks will move freely between the warm and cool areas of the brooder. If they are huddled under the lamplight, the brooder is too cool. If they are huddled far away from the lamplight, the brooder is too warm.

WANT EGGS?
RAISE A LEGHORN

The Leghorn is a beautiful Mediterranean breed known for its reputation as a prolific egg layer. The average Leghorn produces around three hundred large, white eggs per year. As with all breeds, this number tapers off as she ages, but some Leghorns have been known to lay around sixty eggs a year at the age of ten!

Leghorns, as their name would suggest, tend to be leggy and carry an upright, lean posture. The white variety is most common and is still considered a heritage breed, despite its popularity with commercial producers.

The average hen lays around 250 eggs per year. Those that are bred to encourage egg production, like Leghorns (seen here), lay more.

Hen-raised chicks follow Mom and learn the ways of the chicken world.

KEEPING CHICKENS FOR EGGS

The Basics

◈

For the most part, the rule of thumb is this: happy, healthy chickens lay the most eggs. Chicken health, including their level of stress, diet, and contentment, has a direct impact on the quantity and quality of the eggs they lay. This is why the backyard setup creates the best atmosphere for good egg production: it allows the chicken keeper to keep a close eye on the flock, to cater to the individual needs of each chicken, and to be rewarded with abundance.

Store your fresh eggs in dated egg cartons at room temperature until ready to use.

Leghorns have a great feed-to-egg ratio, meaning they lay a lot of eggs in comparison to the amount of feed they eat.

THE BACKYARD SETUP GIVES YOU CONTROL

When you raise a flock of chickens, you soon learn that what you put into your flock is directly related to what your flock produces. I've found that the best strategy is to provide your flock with the most natural living situation that's possible: fresh air, space to roam, and a varied diet through free-ranging. Then, enhancing that lifestyle with quality feed, fresh water, the occasional treat, supplements of your choice, a safe and clean place to live, protected roosts, comfortable nest boxes, and prompt medical care provides a winning combination so that you will get the most from your flock.

The bottom line: spoil your chickens. Spoiled chickens are happy chickens, and happy chickens lay numerous, healthy, and delicious eggs. I derive great pleasure from knowing that I'm taking good care of our chickens—treating them not as a commodity but as living creatures that produce something of value.

Chickens are hardy little animals that can tolerate a lot—but why should they? The easier you make their lives, and the more you create a living environment that complies with their instinctual habits, the more productive your flock will become.

DIET

Biologically, chickens were not necessarily meant to lay as many eggs as we expect from domesticated birds. The domestic chicken is a

If you have early morning layers, your eggs might stay cleaner if you collect twice a day—once in the morning and once in the late afternoon. Never disturb a hen while she is laying an egg, because this can cause injury to the bird.

descendant of the jungle fowl. In the wild, jungle fowl only lay during the breeding season, for a total of about twelve eggs. Over the years, though, chicken breeders have selectively bred chickens to have the capacity to lay as many as 300 eggs per year. However, laying that many eggs requires a diet higher in protein and calcium to support egg production.

Chickens are omnivores, which means their systems are designed to process both plant and animal proteins. In nature, chickens are scavengers—opportunists whose diet consists mainly of foraged plants, berries, nuts, insects, and worms, and occasionally a small rodent, snake, or frog. This varied diet provides a good assortment of plant-based nutrition and complete proteins. Complete protein is especially important to a chicken's health. If deprived of protein for long periods of time, chickens will often resort to feather plucking, egg eating, and even cannibalism.

In the backyard setting, protein is usually provided in the form of soybeans added to a commercial feed. For laying hens, a protein quantity of 16 to 18 percent is best. Additional protein can be provided to your flock through mealworms or by cooking eggs and feeding them back to your chickens. Raw eggs should not be fed, as this can teach your chickens the habit of egg eating.

EGG LABELS: YOU MAY NEED YOUR DICTIONARY TO BUY EGGS

When I was a kid, eggs were eggs. Our local grocery store sold one brand of large white eggs, and the only thing we knew about the origins of these eggs was that they came from a chicken—or so we hoped. We had no idea of how the chicken lived, what it ate, how healthy it was, and so on. Over the years, though, people have become more interested in where their food comes from. And as a result, the egg department of your grocery store has probably expanded since the 1980s.

When you buy a dozen eggs today, you now have choices, decisions to make, and ethical dilemmas. Words like "Cage Free," "Free Range," "Pasture Raised," "Organic," and "Vegetarian Fed" probably didn't enter into your daily grocery considerations until a few years ago. Once a simple task, buying eggs has now become confusing. What's best for you? What's best for the chicken? Did it live a happy life while laying its eggs? Am I paying extra for something that I may or may not agree with?

Let's explore some of these labels.

- **Cage-free** This means what it says—sort of. The chickens are not kept in cages, but under this label it is acceptable for chickens to live in very cramped warehouses with thousands and thousands of birds packed into tight quarters. The label doesn't always imply this, but neither does it guarantee humane conditions.

- **Free-range** This is a loose term for chickens who live with access to the outdoors. This can mean many things. I consider our own chickens free-range because they live in a coop but have access to the outdoors via their run and to our 14 acres when we are home for the day. Our chickens enjoy grass, insects, and sunshine; however, they are supplemented with a nonorganic commercial feed. These conditions are one form of free-range.

When we are home for the day and can keep an eye on our flock, our chickens enjoy exploring the natural setting of our 14-acre farm. If we have errands to run, we keep the flock locked in their coop and run area to prevent the chickens from walking in the road or being attacked by predators.

But free-range can also mean that thousands of chickens are packed into a warehouse with one small door that leads to a small outdoor run. For many of these chickens, life looks very similar to that of a cage-free chicken.

• **Pasture raised** These chickens are raised almost completely on grass, insects, and whatever else they can find in nature. Often, chickens on pasture must be rotated to different areas to provide new pasture. This system requires large plots of land and an increased risk of predator attacks. Commercial feed may be supplemented in small amounts or during the winter months.

Pasture-raised eggs are usually the most expensive because the chickens are not fed a steady diet of commercial feed and therefore don't lay as many eggs. Additives, such as calcium and protein, in commercial feed blends optimize egg production. The upside is that chickens raised on pasture live the most natural life possible for a domestic chicken. The eggs, as a result, are also the most healthful, packed with natural nutrients.

- **Organic eggs**   These are from chickens that are fed an organic feed. But this label alone implies nothing about the living situation of the chicken. Chickens who live on an organic-certified farm and are fed organic feed are considered to lay organic eggs, but that doesn't mean they don't live in cages.

- **Vegetarian fed**   Chickens are natural omnivores, so "vegetarian" is not necessarily a healthier option for them. However, when chickens are vegetarian, you know that they are not consuming chicken by-products and other questionable ingredients—if that is of concern to you. But by a strict definition of "vegetarian fed," the chickens may be kept locked up in a warehouse or a caged situation to ensure that they are not scavenging for insects, frogs, and so on.

- **No label**   Many commercial egg producers keep chickens in cages. These eggs are usually the cheapest because the production of eggs has been streamlined into a machine-like system. The cage containment makes feeding, waste removal, and egg collection efficient and cost-effective. However, keeping chickens in small cages is not something that companies usually like to advertise, so if the egg carton is unlabeled and the price is low, these are most likely the kind of eggs you are buying.

Raising your own backyard chickens frees you from worrying about all this. You know what your chickens eat and how they're raised, and you don't have to trust carton labels to get you through breakfast.

PASTURE RAISED VS. COMMERCIALLY PRODUCED

An article published by *Mother Earth News* in 2008 compared eggs from chickens raised on pasture versus those raised commercially that were then tested for nutritional content. The eggs of chickens raised on pasture contained:

- One-third less cholesterol
- One-fourth less saturated fat
- Two-thirds more vitamin A
- Two times more omega-3 fatty acids
- Three times more vitamin E
- Seven times more beta-carotene
- Four to six times more vitamin D

The increase in nutrients likely comes from a more natural lifestyle for the chickens. Pasture-raised chickens are free to eat healthy, nutrient-dense weeds, grass, and insects; they live cleaner lives and absorb sunshine and fresh air on a daily basis.

GRADING EGGS

The U.S. Department of Agriculture (USDA) has set up a system to help consumers and egg sellers label eggs in a manner that ensures a *relatively* consistent product. This system takes shape, size, and quality, but not shell color, into consideration. Weight is categorized according to a different scale. Grading helps manufacturers package eggs of similar quality, so that consumers are familiar with what they are buying and know what to expect. This is especially helpful when you are choosing eggs for recipes.

Under the USDA system, eggs are graded into three categories: AA, A, and B. A good way to understand this system is to think of AA classification as blue ribbon winners, A class as second place, and B class as last place.

Exterior grading The first step in grading an egg is to simply look at it. The shell is checked for cracks, blemishes, uniform thickness, and shape. Cracked eggs should not be eaten or sold.

Interior grading When grading eggs, both the exterior and the interior are considered. You may be wondering, "How is the interior of an egg graded without cracking open the egg?" The interior of the egg is viewed by candling—the same process of shining a light through the eggshell to determine the development of a baby chick. In the candling process, three things are checked for quality: air cell depth, the albumen (or white), and the yolk.

- **Air cell depth** The air cell is a layer of air that rests between the outer and inner membranes. At the rounded end of an egg, the layer opens up into a larger pocket of air. The size of this pocket can determine how old an egg is. As an egg ages, moisture evaporates through the pores in the eggshell, so as the matter inside shrinks, the air pocket gets larger.

Marigold is often an added ingredient in high-end layer feed. The flower increases the yellow pigment of yolks.

- **Albumen** The white of the egg is checked for thickness, viscosity, clarity, and the presence of impurities. A good indication of a firm albumen is little movement of the yolk when the egg is rotated or tilted.

- **Yolk** The yolk is checked for size, shape, quality of the outline, and the presence of blood spots or other impurities.

Quality factor	AA quality	A quality	B quality	Inedible
Air cell	⅛″ or less in depth	³⁄₁₆″ or less	More than ³⁄₁₆″	Doesn't apply
White	Clean, firm	Clean, may be reasonably firm	Clean, may be weak and watery	Doesn't apply
Yolk	Outline clearly visible	Outline may be fairly well defined	Outline slightly defined	Doesn't apply
Spots (blood or meat)	None	None	Spots aggregating not more than ⅛″ diameter	Spots aggregating not more than ⅛″ diameter

EGG SIZES

Size category (per dozen)	Minimum weight
Small	18 oz.
Medium	21 oz.
Large	24 oz.
Extra Large	27 oz.
Jumbo	30 oz.

DIET AND YOLK COLOR

Backyard chicken keepers pride themselves on the rich, dark, yolk color that is classic from homegrown eggs. Egg yolks get their sunny color from a carotenoid called xanthophyll. The xanthophyll pigment is also responsible for the brilliant color of autumn leaves.

Dark leafy greens are the perfect source of xanthophyll. As it turns out, many of the same foods that contain xanthophyll are also high in nutrition, which is why we attribute dark yellow yolks to healthy eggs.

Adding alfalfa or marigolds to your chicken's diet can also increase yellow pigment.

The difference between store-bought egg yolks and yolks from backyard or small-farm free-range poultry is obvious.

While uniformity is important in egg quality, customers who buy from backyard chicken keepers enjoy a colorful and interesting array of colors and sizes.

Applying the Grading System

Grade AA eggs are of the highest quality on both the inside and the outside. The shell is clean and blemish-free, uniform in thickness, and free of cracks. The egg should have a nice oval shape, with one end slightly pointed and the other end rounded. The interior of the egg is also of the highest quality. Both the yolk and the albumen are firm and free from any impurities.

Grade A eggs have the same quality shell, are free of cracks and blemishes, and are nicely shaped. However, the interior quality is slightly less than that of Grade AA eggs.

Grade B eggs are noticeably inferior on both the outside and the inside. The shell might have blemishes or be oddly shaped, and the interior quality is even less than that of Grade A eggs. I keep our Grade B eggs for our own use in baking or to be scrambled. They are still edible, but should not be sold. Large producers send Grade B eggs to be used as powdered egg or liquid egg products.

Sizing Eggs

When you buy a dozen large eggs at the supermarket, the egg you crack open for your recipe was not weighed individually and determined to be a large egg on its own. Instead, eggs are weighed on an average per dozen. However, when selling eggs, it's important to keep in mind that your customer is looking for uniformity.

TIPS FOR CLEANING AND STORING EGGS

• If you have a stubborn stain on an egg, try dipping the egg in warm vinegar. The stain will loosen and can be scrubbed off easily.

• Apply a light coat of cooking oil to clean eggshells. It not only gives the eggs a pretty sheen, but it will also help prolong the shelf life by sealing some of the pores on the eggshell.

• Store eggs pointed-side down for longer freshness. With the air cell on top, it will slow the exchange of moisture through the eggshell.

Recently washed eggs. Keep your egg-washing sponge separate from your dish-washing sponge to help prevent cross-contamination.

HYGIENE AND REFRIGERATION

When you have a farm, the barrier between the outdoors and the indoors is easy to breach. One year, two of our doe goats gave birth in early spring when the weather was still brutal. Both does rejected their set of twins. To keep the babies from freezing, we brought them inside, where we bottle-fed them and kept them warm until the weather finally broke.

This represents something of an emergency situation, because, in my opinion, the farm is the farm and the house is the house, and there is a civilized line that I try not to cross. In the same spirit, I generally maintain a barrier from the coop to the kitchen, though this barrier is broken each time you bring eggs in from a coop where chickens live. Chicken keepers are strongly opinionated when it comes to the debate for or against washing eggs. The final decision and

routine in each household should be made with knowledge of the risks and benefits.

It's important to understand that an egg is not a closed system within an airtight shell. The eggshell is made of a porous material and is covered in microscopic pores that allow oxygen and humidity to reach the potential chick inside. When a chicken lays an egg, she coats the eggshell in a bacteria-resistant layer called the *bloom*. This invisible layer limits bacteria from passing through the pores in the eggshell and protects the chick.

It is a fact that chickens carry diseases that can be passed on to humans. The most well known is salmonella, a bacterial infection that can be passed via infected chicken droppings. Symptoms of a salmonella infection in your flock include dark combs and wattles, weakness,

Wire baskets make great egg collecting and storage containers, allowing airflow to keep eggs fresher for longer periods.

yellow or greenish diarrhea, increased thirst, and loss of appetite. These symptoms are similar to other chicken illnesses as well. Chickens can also carry salmonella while showing no symptoms.

When a healthy human adult gets salmonella, it is usually an unpleasant experience with symptoms similar to food poisoning. (Often "food poisoning" is, in fact, a salmonella infection.) In rare instances, or with people with lowered immune systems such as the young or the elderly, salmonella can be a life-threatening disease.

In addition to salmonella, there are other bacteria and parasites that can be passed to humans via chickens. In the wild, chickens scratch and peck as they merrily move through their natural

Clean eggs can be stored in bowls in the refrigerator. Our turkey eggs (shown here) often need a good scrub, as turkeys are ground layers.

surroundings. They defecate and move on. They take dust baths in clean patches of dirt, preen their feathers regularly, and are fairly clean animals. But when placed in a coop setting, things change. The confinement of chickens changes the way bacteria presents itself among your flock. No matter how clean your chicken coop is, bacteria are present—in the bedding, in the nest boxes, on your chicken's feet, and on the birds themselves. It's a rule in our house that if you touch the chickens or enter the coop, you must wash your hands before eating or touching your mouth or eyes. This rule also carries over to govern how we treat our eggs before eating them. But many chicken keepers are considerably more relaxed about their handling of eggs, and they rarely, if ever, get sick.

The egg-cleaning debate is so controversial that each state has its own laws and opinions as to what makes for a safe egg. I encourage you to do your own research and consider how your family prepares eggs. Do you make your own homemade mayonnaise with raw eggs? Do you eat fried eggs sunny-side up? Do you make a habit of licking the beaters of their raw cake batter and cookie dough? If yes, then maybe you should be very diligent when it comes to egg-cleaning practices.

For many people, if an egg appears clean, with no visual marks of debris, or droppings, they don't wash the egg before eating. I, however, advise you to err on the side of caution. I assume that anything that's been in my chicken coop has some sort of chicken-related bacteria. No matter how clean your coop appears to be, it is still a place where animals are kept. Birds are especially good at spreading bacteria around because of the nature of their movements. Birds fly, of course, so their bacteria can be transferred to all surfaces. Dried droppings break down into powdered form and become airborne when birds flap their feathers. Dirty chicken feet scratch around in the nest boxes before the egg is laid. So even if dirt is not visible, bacteria could possibly come into contact with a freshly laid egg. Is it enough to make you sick? The chance is there, though probably low, especially if you plan to cook the egg thoroughly.

When you crack an egg, no matter how careful you are, a bit of the inside of the egg comes into contact with the outside of the eggshell. I frequently drop a bit of shell into whatever I'm cracking an egg into and then have to fish out the piece afterward.

My routine is this: when I bring eggs in from the coop, I take them out of the basket and place them in recycled egg cartons labeled with the date. These cartons have not been washed, since they will be holding dirty eggs.

The eggs can be left out on your counter until you wash them and will stay fresh for several weeks. We leave our eggs out and wash them as we need them. Unrefrigerated eggs seem to taste better, but you can feel free to wash them all ahead of time so they are ready to use.

We wash our eggs in warm soapy water with a scrubby sponge dedicated to egg washing. Warm water is important because cold water will shrink the pores, increasing the possibility of drawing bacteria into the eggshell. Do not soak eggs, because this allows bacteria to pass freely from the soaking water into the shell.

Clean eggs should not be stored in dirty cartons. If I wash extra eggs meant for use later, I store them in a clean bowl or in a clean ceramic egg dish in the refrigerator.

If you are selling eggs, check with your county extension for state regulations on cleaning eggs intended for commercial sale.

Remember, too, that clean eggs start in the coop. The cleaner your coop and nest boxes, the cleaner your eggs will be. A dirty coop and run cause dirty chicken feet. They will carry that debris with them to the nest box, and as they step on previously laid eggs and the nesting material, the eggshells will get soiled. Filthy eggshells are not appetizing and are difficult to wash. I try to keep the nesting area as clean as possible. If an egg is too dirty, I discard it because it's not worth the hassle. Finally, I line our nest boxes with fresh pine chips each day.

Collect Eggs Daily or Twice Daily

Being diligent about egg collection helps keep eggs clean. Leaving eggs in the nest box for more than 24 hours increases the chance of an egg being broken, because as eggs pile up, the chickens have less room to walk as they enter to make their deposits. A broken egg causes a mess in your box, and dried egg is very difficult to clean off of eggshells. And a broken egg can also lead to egg eating—a habit that begins when a chicken learns that her eggs are delicious and then begins to break them open deliberately. This behavior is hard to correct and can lead to you finding empty nest boxes each day.

Depending on your chicken's routine, collecting eggs once or twice a day should help keep your boxes cleaner. I collect around 7:00 in the

I like to keep a bag of pine shavings for refreshing the nest boxes each day. The bottom of these metal boxes remove for easy cleanup, then I throw a few handfuls of fresh shavings in each morning.

evening. We have a lot of chickens and our boxes are pretty busy during the day, so I don't like to disturb them if they're in the middle of laying eggs. Our chickens don't start laying until late morning or early afternoon. I find that 7:00 p.m. is a good time because most of the chickens have laid their egg for the day and none of them has yet started roosting for the evening. Our chickens don't roost in the nest boxes but on the bar that is in front of the boxes. If they're already perched up there for the night, egg collection can be cumbersome.

Discourage Chickens from Roosting in Nest Boxes at Night

This brings me to my next point. If you don't provide enough roosting spots, your chickens will sleep in their nest boxes. They will defecate in the boxes all night and soil the bedding, causing dirty eggshells.

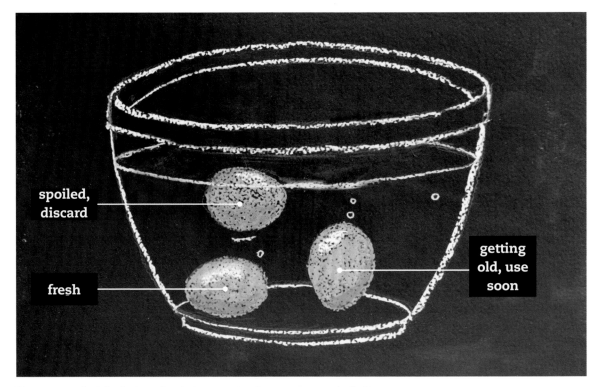

spoiled, discard

fresh

getting old, use soon

If you want to check for freshness but don't want to break an egg, you can do the float test.

Checking for Spoiled Eggs

Farm-fresh eggs don't have an expiration date, so if you end up keeping the eggs for an extended period of time, you might wonder whether they are still okay to eat. There are three ways to check the freshness of an egg.

- **Smell test** When I crack an egg I always smell it before I add it to a recipe. You should smell nothing, other than perhaps a slight chalky scent. No distinct odor should be present.

- **Check the yolk** Many times, if you crack an older egg the yolk will break easily. The egg might still be edible (use the smell test), but if the rest of the dozen are around the same age, use that dozen quickly or perform the float test (above).

- **Float test** If you don't want to break an egg, you can do the float test. Wash the egg and drop it into a bowl of room-temperature water. If the egg completely floats, it's old and should be discarded. Sometimes eggs will stand on end—this is fine, but they should be eaten quickly because this is a sign that they are getting old. As an egg ages, air passes through the eggshell and the egg loses volume. A spoiled egg will have lost enough weight that it will float in water.

GROUND LAYERS

Eggs from ducks, geese, turkeys, and ground-laying birds are more difficult to keep clean. Ducks like to be wet, and they track moisture with them everywhere they go.

Even with your best efforts, it is more difficult to collect pristine duck, geese, and turkey eggs right from the coop, but here are a few things that will help.

• Provide a ground-level nesting box. It can be elevated a few inches, and they appreciate a nest with a roof. Refresh nesting material daily, and keep the box as far away from the water source as possible.

• Provide enough boxes. If you have a mixed flock of ground layers, the different breeds can become territorial toward a specific nest. Our geese have kicked the turkeys out of their nesting boxes and now hiss if anyone comes near their newly claimed nest. We then had to provide the turkeys with a new box to stop them from laying willy-nilly on the coop floor.

Providing a ground-level laying box will keep turkeys, geese, and ducks from laying eggs all over the yard and coop floor. A larger box like this also helps keep eggs from ground-layers cleaner, as well as appealing to the instincts of nesting fowl to lay in the same spot each day.

BREEDING CHICKENS FOR EGG COLOR AND OTHER CHARACTERISTICS

Breeding chickens can be a fascinating backyard hobby and can even result in a productive business opportunity. In its simplest form, it is a tool to perpetuate your flock.

Poultry eggs come in a great variety of colors. When selecting birds to breed for egg color, you need an understanding of their genetic makeup.

As CHICKENS AGE, they lay fewer and fewer eggs, and breeding can ensure that you always have young birds freshening your flock and your egg supply. In its most complicated form, breeding can be an intricate system that analyzes genetics using scientific information to predict the traits of offspring for a desired outcome.

Breeding, for me, is the most interesting aspect of chicken keeping. It's amazing to witness new life and know that you had a hand in steering how that life was created.

One common misconception is that a hen needs a rooster in order to lay eggs. A hen will lay eggs with or without a rooster, although some chicken keepers swear that a rooster will encourage more egg production.

THE BASICS OF BIRD REPRODUCTION AND BREEDING

There is often a misconception among those new to chicken keeping that you need a rooster for a hen to lay eggs. This is not the case. A hen will lay the same number of eggs with or without a rooster present. (Although some chicken keepers swear that a rooster will encourage more egg production.)

Before reproduction age, a female chicken is called a *pullet*, but once she starts laying eggs she is fertile and considered a hen. A hen is capable of laying fertile eggs that can hatch into chicks—provided they are fertilized by a male. If the eggs are not fertilized, the hen will still lay eggs, but they will be infertile eggs that cannot be incubated.

For a hen to lay a *fertile* egg she must mate with a rooster before the egg is formed in the oviduct. As we discussed in Chapter 2, during fertilization, the sperm enters the egg in the infundibulum, or funnel, and appears as the germinal disk on the yolk.

A male chicken before reproduction age is called a *cockerel*. It can be assumed that a cockerel becomes sexually mature (*fertile*) once his comb turns a brilliant red, he starts crowing, or there is visual confirmation that he is mating with hens.

In the very beginning, roosters and hens may need a little extra time while their reproductive tendencies get sorted out. A hen will often lay small, yolkless, or misshapen eggs in early puberty. Most likely, her first few eggs will not result in viable chicks.

Likewise, a rooster takes a while to master his love dance. You may see him attempting to mount hens backward, or he may fall off mid "dance." His crow may also seem somewhat pathetic at first, as he works on his signature song. Once mastered, a rooster's crow stays pretty consistent. If you have more than one rooster, you may find that you can tell which rooster is sounding off by the tone of his crow.

(above) This Easter Egger cockerel shares the rose comb and beard of its Araucana ancestors.

The first few eggs that a young chicken lays are usually odd-shaped, small, or yolkless. Here is an example of a typical French Black Copper Marans egg and a smaller egg called a "wind egg." Allow your hen's system to become regular before collecting eggs for incubation.

First-generation Olive Egger eggs going into the incubator. This is a mix of a Black Copper Marans egg (dark brown) fertilized by an Easter Egger rooster, and two Easter Egger eggs (bluish green) fertilized by a Black Copper rooster.

Viability of Fertile Eggs

Once a fertile egg is laid, it will stay viable—able to be incubated to the chick stage—for approximately 10 days. The older the egg, the less viable is becomes. This window of stored fertility within the egg makes it possible for breeders to ship hatching eggs across the country for people to hatch out chicks from a distant flock. This natural convenience opens up breeding possibilities.

When collecting fertile eggs intended for hatching, I always date the egg with a pencil and keep it in a cool area, below 70°F. Some people refrigerate the eggs, but I've never found this necessary. Try to collect clean-from-the-nest eggs. It is possible to wash hatching eggs, but this removes the protective bloom, exposing the embryo to additional bacteria passing through the pores of the egg.

Our incubator holds seven eggs. I try to fill the incubator to maximize the number of chicks I can get from that run. Sometimes I only have two or three fertile hens that I'm collecting from, and at one egg a day or less per hen, it will take me a few days to collect enough eggs to fill the incubator.

In nature, a hen will lay a clutch of eggs over the course of a week or two. When she has a substantial clutch, she will sit on all the eggs at once, which ensures that all the chicks will hatch out roughly around the same time (rather than one each day, as they were laid). This is important to the survival of the nest, because once chicks hatch, they want to explore their surroundings. The hen would have a hard time chasing after hatched chicks while continuing to sit on those still needing incubation.

Some roosters understand the mating process very quickly, and others take time. If he is a good example of the breed, be patient with him. The offspring will be worth it.

When a rooster mates with a hen, she will remain fertile anywhere from 7 to 14 days, which means that the eggs she lays within this time could potentially become chicks with incubation.

DOUBLE YOLK

When choosing eggs to collect for incubation, you may be tempted to try your luck with the largest eggs you can find in your nest boxes. Many times gigantic eggs are a double yolk. We have an Easter Egger who consistently lays a double yolk, and while it's great for a hearty breakfast, twin chicks hardly ever survive to hatch day. The chicks compete for nutrients and egg space, and if they do make it, they often need outside assistance in order to complete the hatch, which can be dangerous to the chicks.

Olive Egger chicks can come in a variety of colors. These newly hatched chicks have yet to fluff out.

Second-generation Olive Egger eggs in the incubator. These Olive Egger eggs were fertilized by a Black Copper rooster to deepen the olive color of the next generation.

Breeding Clean Lines

If your hen has been exposed to other roosters in your flock and you want her to breed "clean lines" with an exclusive rooster, she should be separated from the undesired roosters for at least 3 weeks. You can keep her with the rooster that you want to hatch chicks with or separate her completely. If you separate her, you can check her eggs for a germinal disk. Once she lays infertile eggs (without a germinal disk), you know she is clean. I try to get a visual confirmation that a hen has mated with the desired rooster, and then I will collect eggs for incubation.

One rooster is capable of inseminating approximately 10 hens consistently, but a tenacious rooster may be able to handle a larger harem.

CHOOSING BIRDS TO BREED

Regardless of the reason you're breeding a particular pair of chickens, it's important to know that you are perpetuating healthy birds. You want to choose robust, lively birds with shining plumage, good body formation, symmetrical features, and no obvious genetic defaults such as cross bill. You want birds with an overall good appearance.

Once the birds have passed a basic, visual health screen, it's then time to research the breed or the end result that you'd like to accomplish with the offspring. If you are breeding chickens that are mainly kept for egg color, there are two things to consider when choosing parent birds.

1. If it's a breed recognized by the American Poultry Association, then the Standard of Perfection should be considered.

2. The unique egg color and laying habits of your chosen hen.

Breeding Resources

If you're interested in breeding chickens, there are two organizations that you should become familiar with.

- The **American Poultry Association (APA)** monitors the breed Standard of Perfection, established by the APA in 1873, and keeps track of recognized breeds and colors.

- **The Livestock Conservancy (TLC)** monitors the status of heritage breeds, encouraging breeders to get involved in hatching rare breeds to prevent extinction. As described by the TLC:

  Before the industrialization of food production, people raised many different breeds of livestock for different purposes. Each breed was painstakingly created by devoted breeders who had a specific goal in mind with each generation. These breeds were hearty, long-lived, and well adapted to outdoor production in various climates, and they provided an important source of protein to the growing population of the country until the mid-twentieth century.

Because each breed had a specific purpose, people interested in buying and raising chickens had many options to fit their needs. For example, those who wanted to concentrate on meat birds searched for breeds such as the Jersey Giant, the Buff Orpington, or the Brahma. Those who wanted egg layers raised Leghorns, Rhode Island Reds, or Sussex breeds. Those who wanted both virtues could choose a "dual-purpose breed" like Wyandottes, Australorp, or Plymouth Rocks, which are good layers but grow to a nice table size as well.

Chicken breeds could also be chosen for climate tolerance, temperament, egg color, taste, or simply for the way they looked. With over 65 recognized breeds, there was the perfect chicken for every need.

The Black Copper Marans cock is an impressive bird. gengirl/Shutterstock

When breeding, it's important to know that you are perpetuating healthy birds. Choose lively birds with shiny plumage, symmetrical features, and no obvious genetic defaults like crossed beaks. Welsummer hens, like this, lay dark eggs with even darker spots.

Once the food industry moved into factories, however, farmers concentrated on a few highly productive breeds and hybrid chickens. At the same time, people moved toward *shopping* for food rather than raising their own. As a result, many of the backyard heritage breeds became less popular and some disappeared altogether. According to the TLC:

> With the industrialization of chickens, many breeds were sidelined in preference for a few rapidly growing hybrids. The Livestock Conservancy now lists over three dozen breeds of chickens in danger of extinction. Extinction of a breed would mean the irrevocable loss of the genetic resources and options it embodies.

So what does all this mean to a person who wants to start breeding chickens?

One of the main goals of the Standard of Perfection was to ensure that breeders had a list of checks and balances to maintain in breeding practice. This ensured that a Marans chicken, for example, continued to look and perform like a Marans over the generations. Without a standard, a breeder could raise any chicken and say it was a Marans. With a standard in place, people could then become familiar with the breeds and know what to expect.

To a novice, these standards can get quite confusing and are best applied through breed exposure and seeing the qualities/disqualifications in examples on an actual bird. It's difficult, for example, to know the difference between the color described as "copper" from, say, "gold" without seeing feather examples of each. Serious breeders spend years perfecting the traits of their flock, resulting in some of the most beautiful chickens you will ever see.

When you decide to breed chickens, you take part in a silent responsibility to perpetuate birds that are good examples of the breed name. This keeps the integrity of the breed intact. Some people are very passionate about this subject and express frustration with "casual breeders." In my opinion, breed integrity should be the ultimate goal, but high standards should not discourage new people from participating in breeding programs or from personally perpetuating their own home flock.

If possible, do what you can for the breed. Be willing to pay a little extra and get your parent stock from a reputable breeder. Learn all that you can from that breeder, asking about the lines and their breeding practices. Most breeders will be willing to share any tips they have about their line, because they have spent many years perfecting it and want to see healthy birds continue.

INBREEDING, LINE BREEDING, OUTCROSSING, AND CROSSBREEDING

So you have a rooster and you have a hen and last spring you mated them and made more chickens. Now what?

One of the most common questions I get asked is, "Can I breed birds that are related to each other?" This is usually a "second or third season" question. When people have purchased chicks, they have one rooster and would like to perpetuate their flock. However, at this point, the rooster is related to all of last year's offspring. So unless you purchase new, unrelated birds each year, inbreeding is inevitable.

Unfortunately, there is not an easy "yes" or "no" answer to this question. A true answer to this dilemma requires a complicated lesson in genetic diversity. It also requires close monitoring of breed traits over successive generations and impeccable records, and most of all, years of experience or an excellent mentor of the breed. Entire books can be written on this subject, but for a beginner's lesson I'll break down the four ways to breed a chicken: inbreeding, line breeding, outcrossing, and crossbreeding.

Inbreeding

Inbreeding is the practice of breeding two animals that are closely related—for example, parent to offspring, or brother to sister.

In the human world, this term has a very negative meaning, implying incestuous relationships not acceptable by cultural norms. However, in the animal world, where moral and ethical issues dealing with unions are not involved, inbreeding is neither good nor bad. It is a circumstance that can be used as a tool by breeders to intensify the genes.

Without getting too technical, I'll simplify the concept. If you have animals with good genes, inbreeding intensifies those good genes.

It can ensure quality traits in offspring. Continued inbreeding with each successive generation creates birds that are closer and closer to being identical. For breeders, this creates a line of birds that will reproduce predictably, which is a good thing if you want to offer them potential customers or create successive show birds with the desired qualities.

Inbreeding gets risky when the birds have bad genes, because inbreeding intensifies these traits as well, usually in the form of undesirable recessive genes or disfigurements. This is because healthy-looking chickens may be carrying an undesirable recessive gene in their DNA. With two unrelated parents, the chances of this gene rearing its ugly head are rare. However, if you mate two birds with the same bad recessive gene, the chances of that recessive gene manifesting in the offspring is greatly increased—potentially resulting in genetic disfigurement. Each time you mate those birds, or the offspring of those birds within the line, the disfigurement becomes more and more likely to surface.

Another problem is that breeding closely related birds generation after generation often leads to infertile offspring. Thus, inbreeding is of limited use.

Line Breeding

Although there is no agreed-upon definition of the term *line breeding*, it usually refers to less concentrated forms of inbreeding—for example, uncle to niece, half siblings, or cousins. (If the relationship of a mating pair is more than five levels removed, they are not "related" enough to be considered inbred.)

Line breeding gives you a similar result as inbreeding, but with less intensification of the genes. So the offspring are not quite as identical, and the risk for genetic mutation is not quite as high.

This is an example of an Olive Egger hen. She shares the body type of her Easter Egger mother, but has coloring similar to that of her Black Copper Marans father.

Outcrossing

For the most part, outcrossing means introducing new genetic information via a new line of birds within the same breed. In other words, it means bringing in fresh, unrelated birds from either another breeder or from a line that you've kept separate in your own stock. Outcrossing, again, is neither good nor bad but can be used as a tool. Just as with inbreeding and line breeding, it can be a useful practice but comes with risks.

Outcrossing your line introduces new genetic information into the offspring. If these new birds have good genetic information to share, then there can be much benefit. If your existing birds have a trait that is undesirable according to the breed standard, new birds might be able to improve this trait in the offspring. Outcrossing also freshens the line and can help with infertility

and other issues caused by overbreeding within the same line.

However, there are risks that come with outcrossing. The first is that these new birds may have undesirable genes that may not have been present in your existing line but that will become part of the new offspring's DNA. Outcrossing also makes the offspring less identical genetically, so it will be harder to predict what the new generation will look like.

Crossbreeding

Crossbreeding is the practice of breeding two different breeds together to get a desired result. One of the most interesting crossbreeds is the Olive Egger, which, as the name suggests, is any nonrecognized breed of chicken that lays olive-toned eggs. This is usually accomplished by breeding a brown egg–laying breed with a green egg–laying breed. And just as when you mix paint colors, the result is an offspring that lays olive-toned eggs. The darker the brown layer's egg color, the deeper the shade of olive that will result in the next generation. More about this below.

Crossbreeding can also be used to bring out certain traits that no longer exist within a breed. For example, some heritage breeds are dropping in numbers because they don't lay enough eggs to proliferate the breed. To combat this, some breeders will cross them with a prolific layer such as a Leghorn to improve the egg-laying capability. Then, they slowly breed the Leghorn genes back out of the flock until the original breed standards are back, but with a tendency to lay more eggs. Crossbreeding can be fascinating, but complicated.

To put these four tools to use in your own flock and breeding practices, you must first learn about the breed you are trying to perpetuate, research the standard guidelines, and weigh the risks and benefits with each breeding practice. Experience is the best teacher, unless you're lucky enough to find someone who can mentor you, who has dealt with the breed for many years and has a good-quality line. Visit poultry shows and talk to exhibitors, judges, breeders, and breed club members. There are many clubs around the country that specialize in specific breeds. Find out about joining one of these clubs. Consider showing some of your chickens to find out how they compare with other lines.

An Olive Egger rooster. Although he won't lay eggs, he carries the Olive Egger egg gene, which he will pass on to his descendants.

A nice example of a Black Copper Marans hen.

BREEDING FOR COLORFUL EGGS

As we've established, first and foremost is the health and genetic quality of the chicken. Regardless of egg color, you want to perpetuate a healthy flock. Also a consistent layer is a good trait to pass down.

Breeding for egg color can complicate the genetic formula even further. Chickens who lay interesting egg colors fall into two categories: those that are recognized as a breed and those that are not. So when selecting birds to breed, you must have an understanding of the genetic makeup of the chicken, along with a consideration for egg color.

Recognized Breeds

One example of a recognized breed that produces unusual egg color is the Cream Legbar. This breed is mostly known for laying beautiful pale blue eggs, but can also lay eggs in shades of green, mint, and olive.

Another interesting breed known for its egg color is the Marans. Marans lay the deepest brown egg of any chicken breed. The eggs can range from a dark brown to chocolate and almost burgundy. As of 2011, three colors of Marans are recognized by the APA: the White Marans, the Wheaten Marans, and, maybe the most interesting of all, the Black Copper Marans.

The Black Copper Marans originated in France. It was domesticated to appeal to French culinary artists in search of the world's most delicious egg. Although the French Black Copper Marans is gaining in popularity and is therefore more available and less expensive, at one time a breeding pair would sell for several hundred dollars. It was said that a French Black Copper Marans omelet could cost up to $50.

Easter Eggers come in many colors and varieties. This hen shows lacing in her feathers.

EARLOBES AND EGG COLOR

If you want to know what color egg a chicken lays, then take a look at her earlobe. It's not 100 percent accurate, but it covers the majority of chicken breeds and will give you a good guess if you have nothing else to go by.

• If the earlobe is red, more than likely you have a brown layer.

• If the earlobe is white, more than likely you have a white layer.

There are a few exceptions to this rule:

• Penedesencas and Empordanesas have white earlobes, but lay a dark brown egg.

• Araucanas, Ameraucanas, and Easter Eggers have red earlobes, but lay a blue or green egg.

• Silkies have a blue earlobe, but lay a white or cream egg.

MARANS EGG CHART

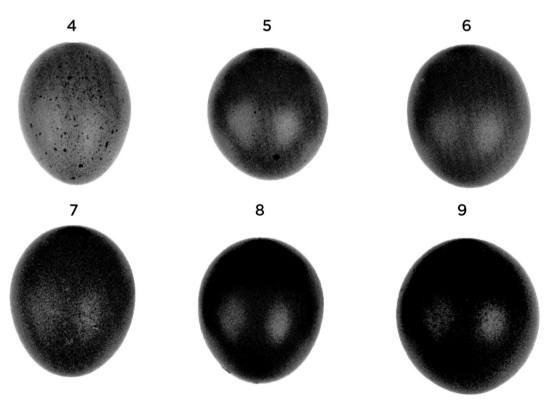

Courtesy Marans Club of America

There are several considerations to keep in mind when breeding chickens such as Marans. Because the breed is recognized, it's important to choose a mating pair that hold close to the breed standard. In addition to this Standard of Perfection, the French Black Copper Marans (FBCM) has an egg color standard. According to the Marans Club, a Black Copper Marans should lay an egg categorized as at least a "4" or darker on the Official Marans Egg Color Chart.

Unfortunately, it is rare that a Marans will breed true to the standard while also having the genes for extremely dark eggs. Plus it seems that the layers with the darkest eggs also lay the fewest eggs. Breeders in the Marans world are working to improve the combination of these qualities, seeking to breed birds that are good in body type, feather color, and more while also laying a consistently dark egg.

Marans also have a cycle to their egg laying. A young hen will lay the darkest eggs in her first year, but as she ages her eggs will lighten slightly in color. She also has a seasonal cycle, in which the first eggs of the year are the darkest, and then lighten as her cycle continues. She will often take a short break from laying, and then start her cycle over again with darker eggs. It seems to be different for each hen, but often the autumn molt marks the stopping point for a FBCM cycle.

Unrecognized Breeds

Two of the most fun chickens to breed are Easter Eggers and Olive Eggers. Because these breeds do not meet a Standard of Perfection, it takes some of the pressure off of the breeder and allows you to concentrate on egg color alone. But no matter why you are breeding chickens, the perpetuation of healthy offspring should always be foremost. The pros and cons of inbreeding and line breeding should also be considered.

THE ARAUCANA, AMERAUCANA, AND EASTER EGGER MYSTERY

Twenty-some years ago, when I first began raising chickens, I started with a pair of Rhode Island Reds and a pair of Araucanas—or so I thought. They came from a feed store that ordered the chicks from a hatchery. The galvanized feed bins where the newly shipped chicks were being brooded were labeled "Straight Run, Rhode Island Red" and "Pullets, Araucana." As chance would have it, one of the pullets turned out to be a rooster.

The "Araucana" chicks came in a variety of colors, with most of them sporting an adorable feather pattern around the eye that was reminiscent of an overdone eyeliner job. They also had black stripes down their back, which reminded me of small chipmunks. Some had fluffy cheeks, some not. Some were dark brown with reddish markings, and others were light tan with cream markings. There were even a few that were light

Easter Egger chick. Notice the dark markings the chipmunk-like stripes behind the eyes.

Easter Egger chickens come in a variety of colors and feather patterns.

gray with even lighter markings. This colorful mix of chicks was fascinating, and what furthered their appeal was the fact that they laid blue eggs. Amazing!

Fascinated by this new chicken, I headed to our local library to find a book on chicken breeds. Araucanas were described as an extremely rare breed—rumpless, tufted, and layers of blue eggs. I was thrilled to have stumbled onto them!

As the chicks grew, I noticed that no tufts appeared. I also noticed that for a supposed "rumpless" chicken, there was a sure presence of a tail. As the hens started laying, my egg basket was filled with eggs that were green, not blue. I decided that the chickens in our backyard were not, in fact, Araucanas, but must be something else. I remembered reading in the book from the library about another breed called the Ameraucana that was similar to the Araucana but lacked the tufts and had a tail.

This helped solve the mystery somewhat, and I believed that the bin must have been mislabeled. But there was still something that confused me: the green eggs! The book said specifically that Ameraucanas also laid blue eggs. Not being able to find any other breeds that came close to the chickens I had, I decided I had some sort of rare Ameraucana whose eggs were on the green side.

(opposite) Another example of an Easter Egger color. The small comb is less impervious to frostbite, making the Easter Egger a favorite breed in colder climates.

This beautiful Olive Egger rooster shows copper-laced gray feathers and a beautiful copper-and-russet collar.

It wasn't until the late 1990s when the Internet came into homes that I first read the term "Easter Egger."

A hatchery website explained that Easter Eggers are in fact a "mutt" breed that can lay green, blue, or even pink eggs. They often resemble Ameraucanas because they share some of the genes of these rarer chickens, but Easter Eggers are also mixed with other breeds of colorful egg layers. It was a bittersweet finding. I was happy to have finally solved the mystery but also disappointed to find that my chickens not only weren't rare, but weren't a breed at all.

I'm not the only one that is confused by this trio of chicken varieties. I still see chick bins mislabeled at feed stores and confused people on chicken forums trying to find out what they have in their coop.

So here's the lowdown on each variety.

- The **Araucana** is a rare, recognized breed that first originated in Chile. As I said before, it is a rumpless, tufted breed that lays blue eggs. The reason the Araucana is so rare is because the same gene that is responsible for the tufting can mutate and be deadly to early chick development.

Easter Eggers come in a variety of colors and feather patterns. They are not a breed at all, but a variety of crossbred chickens carrying genes that allow them to produce a variety of colored eggs.

- The **Ameraucana** is a recognized breed created in America in the 1970s as the result of trying to correct the deadly gene mutation. The result is a bearded breed with muffs and a tail. But it still holds the blue egg-laying ability.

- **Easter Eggers** are not a breed at all, but a variety of crossbred chickens that carry the genes to lay a variety of colored eggs, including blue, green, and pink. Many Easter Eggers will have Ameraucana or Araucana blood in them, which explains some of the similarities in looks, but they are not true to the breeds.

Most hatcheries do not sell true Araucanas or Ameraucanas. They are only available through breeders, so if you're getting chicks from a feed store, they are probably Easter Eggers, regardless of what the label says.

OLIVE EGGERS

Olive Eggers are a fun "breed" whose egg color can vary with each generation. The olive color comes from crossbreeding a brown egg-laying chicken with a green egg-laying chicken. The darker the brown egg color, the deeper the olive egg from the next generation. I've had the best success mating a French Black Copper Marans hen to one of our Easter Egger roosters. The resulting offspring lay a gorgeous dark olive egg. But I've read of many successful crosses mating a Marans rooster to a green egg-laying hen.

Many people have also created a pale egg shade from crossing a light-brown egg-laying hen such as a Wyandotte or Buff Orpington to a green egg layer, resulting in light olive eggs.

The egg color can be further changed by successive generations. If you mate an Olive Egger back to a Marans, the next generation will lay an even deeper shade of olive. This is where the amateur breeder should carefully weigh inbreeding and line breeding. Be aware that if your rooster is already the chick's father, breeding him again with the new generation will be considered inbreeding.

SELECTING EGGS FOR HATCHING

If the chicken that you're interested in breeding is mainly raised for egg production or a certain egg color, then focus on your nest boxes when choosing the eggs you want to incubate and hatch. If you have a small number of chickens, it can be easy to keep track of who is laying what, but even if you have a large number of chickens you will begin to recognize a certain shape, color, and size of egg if you pay close attention. For example, we have a Buff Orpington who always lays an elongated egg. I'm not sure which Buff lays this egg, but she is consistent. When collecting eggs for hatching, I will not choose this egg, thereby eliminating the risk of this trait being passed down to her offspring.

And if you are avoiding eggs with undesirable characteristics, then of course you will be looking for desirable eggs to hatch. If you have a Marans that lays a darker-brown egg than the rest, choose that dark-brown egg to incubate. Hatch the egg that you want to see repeated in generations to come. And choose well-shaped eggs for hatching. Eggs that are small, extremely large, or oddly shaped are often infertile.

HOW EGG COLOR IS FORMED

The genes of a chicken determine what color egg she will lay, and the color is determined in two different stages of the egg's development. The first pigment can be determined in the shell development itself, the second in the development of the cuticle.

Over a period of 15 hours, the cells in the uterus add a calcium carbonate shell over the egg. Once this process is complete, the formation of the cuticle begins. This is also where pigment is added to brown and green eggs.

EGG COLOR BREED GUIDE

Breeds That Lay White Eggs

Leghorn
Ancona
Hamburg

Breeds That Lay Light-Brown Eggs

Faverolles
Naked Neck
Sussex

Breeds That Lay Medium-Brown Eggs

Australorp
Brahma
Cochin

Breeds That Lay Dark-Brown Eggs

Marans
Barnevelder
Welsummer

Breeds That Lay Blue Eggs

Araucana
Ameraucana
Cream Legbars

Breeds That Lay Green Eggs

Easter Eggers
Olive Eggers

Notice the darker pigment distributed by the cuticle formation, which can be scratched off to reveal a lighter eggshell below.

A trio of green Easter Egger eggs. A green egg results from a blue eggshell coated in brown cuticle pigment.

A Welsummer egg with darker-brown spots. Brown eggshells receive their pigments during the formation of the cuticle—in other words, a brown egg is a white egg coated in a brown cuticle.

The color white or blue in an egg is developed during the shell formation process. It is the true color of the eggshell. Brown is distributed as a pigment in the cuticle layer. In other words, a brown egg is a white egg coated in a brown cuticle. A green egg is a blue eggshell coated in a brown cuticle pigment. In some of the darkest brown eggs, the cuticle layer is so thick that the brown color can actually be scrubbed off. This thick cuticle lends extra protection against bacteria entering the egg, which is why darker eggs may stay fresher longer.

EGG COLOR AND CHICKEN BREEDS

For the most part, each breed has a corresponding egg color that all chickens of that breed can be expected to lay. For example, all Rhode Island Reds lay a medium-brown egg, all Leghorns lay a white egg, and so on. The shade of a colored egg may lighten or darken throughout the hen's cycle and lifetime, but brown will stay brown and white will stay white.

HOW TO RAISE OLIVE EGGERS

**Easter Egger
(green eggs)**

**French Black Copper Marans
(chocolate eggs)**

+

=

Olive Eggers!

OTHER POULTRY AND THEIR EGGS

Chickens are by far the most popular backyard fowl.
The popularity of raising chickens is due to the convenient
benefits that a backyard flock provides and the fact
that chickens are easily found for sale each spring.
Chickens are easy to keep, require less work than a dog,
need little to thrive, and produce eggs.

Chickens may be the first birds that come to mind when we think of farm-fresh eggs, but keeping a flock of other types of poultry can be just as easy and rewarding. Duck eggs are gaining in popularity among poultry and food enthusiasts.

This handy photo depicts the relative sizes of (from left) quail, chick, duck, and goose eggs.

MANY PEOPLE DON'T realize that keeping a flock of other types of poultry can be just as easy as raising chickens. Some fowl, such as quail, require much less space; ducks are quieter than chickens, and turkeys are less messy. Larger fowl, including turkeys and geese, require more space and may require special permission from neighborhood associations, but the daily upkeep is usually quite similar.

In this chapter, we will explore keeping ducks, turkeys, geese, guineas, and quail as backyard flocks.

DUCKS

After chickens, ducks are probably the most popular backyard bird. Like chickens, ducks are usually raised for eggs or meat, or kept as pets.

Much of the backyard poultry movement is driven by whatever young birds are offered by the larger feed stores in the spring, and this is what piques the curiosity of people who might not otherwise be interested. People fall in love with the baby birds available in the brooders and realize that keeping backyard poultry is not only possible but also pretty easy. While hovering

A duck's natural habitat is near the water. Swimming encourages proper digestion and feather health in ducks.

around the brooders in the spring, I often hear people in conversation about the prospects of starting their own flock. I've even talked a family or two into it myself.

Often, ducklings are offered alongside chickens at the feed stores. Turkeys are becoming more mainstream as well, but turkeys are large and are usually raised only for meat or to proliferate the breed, which causes many newbies to shy away from that endeavor.

Ducks are easier to raise than chickens in some respects and harder in others. Ducks don't scratch, and because most breeds don't roost, most of the mess is kept to the floor area of the coop. However, ducks LOVE water. They will mix everything with water, splash, bathe, and muck up all the water pans. Where chickens stir up dust, ducks saturate their living space, making cleaning a heavy, wet job.

WHY DOES A DUCK FLOAT?

If you've ever seen a flock of ducks sitting by the waterside preening their feathers, there's more going on than meets the eye. Not only are the ducks arranging and tidying their feathers, but they are also distributing oil over the feathers. The duck collects the oil from a gland at the base of the tail and spreads it over the feathers. This protects the feathers and makes the duck waterproof. The outer oil-coated feathers help keep the downy under-feathers dry and keeps the duck warm in cold water.

(above) The Muscovy duck lays a deliciously rich egg, though not in the same quantities as breeds like the Khaki Campbell.

(right) Khaki Campbell ducklings. Today, ducklings are offered alongside chickens at the feed stores.

Ducks are social creatures and do best when in the company of other ducks.

Ducks will turn any water into an opportunity to splash.

Pekin duck eggs. Duck eggs are slightly larger than chicken eggs and have become increasingly popular in recent years. Foodies are recognizing their subtle differences in appearance, taste, and texture.

It is possible to raise ducks without a large body of water, but they will be much happier if they can splash and swim around. Even something as simple as a kiddie pool will make for a happy flock. If denied a swimming vessel, they will attempt to splash in anything that's available—usually the water dish.

Ducks need water not only to rehydrate but also because they must mix their food with an outside water source in order to swallow. This is where the wetness comes from. Ducks will track water from the water dish to the feed dish and vice versa until the water is brown with food and the food is mucky with water.

Ducks are medium-sized birds, similar to a chicken, and therefore require similar housing space. But because they are less able to defend themselves against predators, extra security in the run and coop is necessary to keep ducks safe. Ducks are a gentler species than chickens, slower and without the chicken's ability to peck in a defensive manner. The duck's bill is smooth and has little biting power, which makes them perfect for children to handle and raise, but it is this very gentleness that makes them more prone to predator attacks.

Ducks do not possess the natural homing instinct that chickens do. As dusk approaches, chickens will instinctually return to their coop and roost for the night, perhaps because the chicken is almost blind in the dark. In the wild, chickens get situated safely in the tree branches before dusk falls and they lose their ability to see. Ducks, on the other hand, can see fairly well at night and often have to be coaxed with food or physically rounded up until they learn to return home by habit.

Ducks are also quieter than chickens, so they work well in neighborhoods that don't allow noisy roosters. In the duck family, the female is the loud one, and she may sound off three or four times a day with a loud, traditional "quack, quack, quack." Still, domestic ducks are relatively quiet birds. The male's vocal range is limited to a hoarse whisper.

More often than not, ducklings are sold as a straight run, meaning you have an equal chance of getting males and females, as it is nearly impossible to sex baby ducklings. Ducks do best when the flock is balanced with one male per female, or with the females outnumbering the males. Sometimes in a duck flock, ducks will pair off in convenient couples, while in other flocks overly zealous males try to mate with all the females in the flock. This can be very stressful to the females; where there is a large body of water, competing males have been known to drown a female.

Some duck breeds lay as many eggs as some chicken breeds do. The Khaki Campbell, for example, is one of the most prolific layers, laying as many as 300 eggs per year.

Duck eggs, which are slightly larger than chicken eggs, have become increasingly popular in recent years. The shell of a duck egg has a smooth, almost velvety texture, somewhat translucent and less chalky than that of a chicken egg. Foodies and niche market areas are recognizing the duck egg's subtle appearance, taste, and textural differences, and many chefs are realizing the potential of duck eggs.

Duck eggs are especially popular in baked goods. The additional protein creates more rise and more fluff in cakes and cookies. When pan-fried, duck eggs can have a slightly firmer texture than that of a chicken egg. It's important not to overcook duck eggs for risk of them going rubbery. An undercooked duck yolk tends to be silkier than that of a chicken egg, because of the additional fat in the yolk.

TERMINOLOGY AND DETAILS

Female bird: Duck

Male bird: Drake

Young bird: Duckling

Gestation: 28 days (Muscovy 33 to 35 days)

Egg size: 3 inches long, 6 inches in circumference at widest point

Egg weight: 2½ ounces

Popular Duck Breeds

The most common backyard duck is the white Pekin, a Chinese duck that is bred for table. This duck lays a creamy white egg that is almost symmetrically oval in shape. The Cayuga is a beautiful breed with iridescent green and black feathers. It lays a greenish-gray egg that ranges to a deep charcoal in color. The Muscovy duck is an interesting bird, with a head covered in red bumps similar to that of a turkey. These bumps are called caruncles.

TURKEYS

I've raised Black Spanish turkeys since 2009, and have found them to be easy and pleasant birds to keep, in many ways a perfect backyard pet. As large birds, however, they require a lot of space.

Turkeys, in general, are much tidier birds than chickens, which will scratch and throw dust and bedding around, or ducks, which get everything wet and saturate the coop and run. Our own turkey pen is always the cleanest coop in the farmyard. Turkeys will occasionally scratch, but not with the same persistence as chickens. When you place a layer of bedding down in a turkey pen, it usually remains undisturbed, and food and water stay in their containers. Turkeys simply don't have the instinct to throw things around.

In character, turkeys have a regal manner—slow, majestic, and deliberate with their movements, and almost stoic in behavior. They are also very loyal to each other. Our tom looks after his hens and follows them around with full feathers on display. He takes his job of protecting his hens extremely seriously and is always diligent in his efforts to come between them and any possible threats. This organization results in a calm, well-mannered flock. He has become a pet over the years, and while we've processed many of his offspring for table, we could never part with our beloved tom.

A good tom makes it his duty to protect his hens and poults.

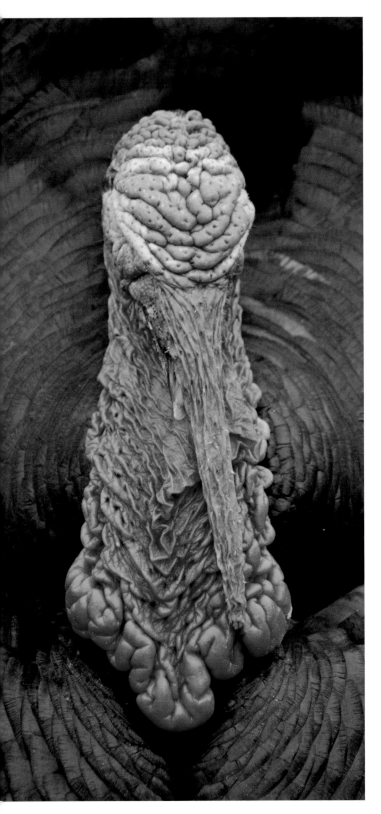

(above) Because most turkey breeds, including this heritage breed Bronze Breasted turkey, are kept for meat, they are not prolific egg layers.

(left) A dominant tom turkey takes his job of protecting his hens seriously.

TURKEYS CHANGE COLORS

Did you know that the head color on a tom turkey will change from pale pink to blue to red depending on his mood? When he is docile and calm, his head color will range from a pale pink or flesh color to sky blue. As he raises his feathers in full display, the blue will deepen and slowly turn red.

Although the eggs from turkey hens are edible, the production is low, and people generally keep these eggs for incubation rather than for breakfast. However, the turkey egg is a most beautiful thing: large and cream colored, with a substantial shell and delicate brown spots. The egg has a satisfying round bulbous end, and is more pointed at the top when compared to a chicken egg.

If you're interested in raising turkeys but can't quite bring yourself to do the butchering, they still make a great addition to the farm as pets. Be sure to socialize with them from a young age. Hand-feeding and handling the poults will make for friendly turkeys. Another use for keeping a pair of heritage turkeys is to sell hatching eggs or poults. This helps continue the availability of otherwise hard-to-find birds, allowing others to enjoy this fabulous species.

These young turkeys have yet to get their adult plumage. Bred mostly for meat, standard breeds are not known for prolific or high-quality eggs.

(above) Black Spanish turkey poults hatching.

(right) Handling and hand-feeding turkey poults will make for friendlier adult birds.

(above) Our turkey pen is always the cleanest coop in the yard. Turkeys simply don't have the instinct to throw food and bedding around.

(left) Turkeys are impressive birds and can double their apparent size by raising the feathers at the base of the shaft.

A turkey egg is a most beautiful thing—a rich cream color with delicate brown spots.

Because turkeys are really only kept for meat, pets, or breeding purposes, the standard breeds are not known for prolific or high-quality eggs. Most of the factory production birds, such as the Broad Breasted White, are incapable of reproducing on their own and must be artificially inseminated. The Broad Breasted White also has a shorter life span. If you want to raise turkeys and continue your flock each year, choose a heritage breed that will mate on its own and produce eggs that can be collected and incubated.

Turkeys aren't always the best mothers among poultry breeds. Often, our hens will sit on nests and then give up halfway through the incubation time for no apparent reason. This is perhaps because turkeys like a lot of privacy when sitting, and if there is too much interruption they will abandon the nest.

GEESE

Geese were originally bred for meat, as guard birds, and for their feathers. But goose meat is a rarity on today's menus, and although goose down is still used, synthetic materials are quickly gaining popularity. On top of that, geese require a lot of space to prosper. Because of these reasons, the need to keep a gaggle of geese on the farm is becoming obsolete, and as a result, many of the geese species are endangered. For most backyard bird keepers, the usefulness of geese lies in their ability to protect the flock, and as pets.

Our geese have become favorite animals on the farm. When they were small, I would take them out with me to weed the garden, where they were content to sit on the edges of my skirt and play with the ruffles while I worked. As they got older, they grew more independent, especially the goose. But our gander will still eat out of our hands and tolerate his feathers being stroked. Had we been more encouraging, I'm sure he would have the personality of a puppy dog.

Because all our birds are free-range, predators can be a problem, and it was here that the geese

Toulouse gander (gray) and Pilgrim goose (white). Geese, like turkeys, tend to lay eggs steadily in the spring and early summer, and then taper off as the warm months progress.

(above) Pilgrim goose. Goose eggs are more practically used for breeding purposes than for eating—one goose egg contains about 400 percent of the daily recommended cholesterol allowance.

(left) Geese require a lot of space to prosper. Nevertheless, ours have become favorite animals on the farm. This gosling is grazing on spring grass.

Other Poultry and Their Eggs

For most backyard bird keepers, the usefulness of geese lies in their ability to protect the flock. Few animals would want to tangle with this fellow.

TERMINOLOGY AND DETAILS

Female bird: Goose

Male bird: Gander

Young bird: Gosling

Gestation: 28 to 35 days

Egg size: 4 inches long, 7 inches in circumference at widest point

Egg weight: 5½ ounces

Goose egg in comparison to a large chicken egg.

proved most valuable. Over the years, we have had difficulties with foxes and other predators, but since adding geese to our flock our attacks have been zero. If anything suspicious comes near our flock, the geese lower their heads and hiss away the threat. The geese can be quite intimidating, and they are especially protective of our ducks, which are particularly vulnerable. The geese and ducks have formed a tight waterfowl bond that works perfectly.

Geese, like turkeys, tend to lay eggs steadily in the spring and early summer, and then taper off as the warm months progress. And also like turkeys, goose eggs are more practically used for breeding purposes than for eating. One goose egg contains about 400 percent of the daily cholesterol allowance. But fortunately, unlike turkeys, geese make wonderful mothers, highly protective of the nest. With my goose, to prevent being bit I had to collect eggs only when she left the nest to drink or eat. To incubate a goose egg, be sure to choose an incubator that will hold a large egg.

Goose eggs are quite beautiful. They are dense and heavy for their size, with a shell that is quite thick and substantial, like a fine layer of porcelain. Because of this, the goose eggshell can be carved into intricate designs using a delicate drill and sanding tools.

Goose eggs are quite beautiful—heavy with a thick and substantial shell.

The French Pearl guinea has beautiful spotted feathers and is an enthusiastic hunter of insects.

GUINEAS

Guineas seem like the soldiers of the farm to me. They march around in their little football shape, sounding off whenever something isn't right in the yard. Guineas are tough little birds.

Guineas can be aggressive, territorial, and food possessive, especially toward roosters. If you plan on raising guineas with chickens, never introduce adult guineas to an adult flock of chickens. More than likely you will have territorial issues, and more than likely the guineas will win the battle. We've found success with raising keets with chicks in the same brooder. The keets seem to grow up thinking they're chickens and will bond with their brooder mates for life. Also, make sure your keets are outnumbered by the chicks, and it doesn't hurt if the chicks are a week or so older than the keets.

Guineas are loud. The females make a loud, repeating two-syllable noise— "*saw wheat, saw wheat*"—and the males make a similar one-syllable sound. The males will also call a short repeated chortle of *kee, kee, kee, kee*, often when

GUINEA CLOWNS

My husband and I have often commented on how much our French Pearl guinea fowl remind us of clowns. As they reach adulthood, their head feathers are shed and replaced by a white, almost "painted" face and head called a helmet. They have large, blue-rimmed eyes and a bright red mound of a comb that reminds us of a red clown wig. In combination with their black-and-white polka-dot bodies, these funny birds have the looks that match their clownlike personalities.

flying or running short distances. But all of these noises are loud. When we are working outside near the guineas and they are voicing, it's hard to carry on a conversation.

Because they are so loud, guineas are often kept as watch animals. But unlike geese, the guineas are not discriminatory about sounding off. They often send loud calls for no particular reason, which may be a problem if noise is a concern in your area.

Guineas can be raised much like chickens. They are of similar size, require similar living situations, and have similar food requirements. However, if you live in an area where roosters are banned because of noise, you might want to rethink guineas, because they will give any chicken a run for his money.

A better reason to raise guineas is for pest control. A guinea's favorite food is ticks, and they love nothing better than gobbling up hundreds of insects a day. Guineas, for this reason, do well as free-range birds. They are somewhat less domesticated than chickens and seem to be savvier when it comes to escaping predator attacks. They are also less tame and harder to train. You might find that your guineas prefer to roost in tree branches at night rather than the security of the coop. They also require more handling to become friendly than do chickens.

Guinea eggs are delicious! We raise French Pearl guineas, which lay a chalky light-brown egg with visible pores, sometimes with a sprinkling of darker brown freckles. Guinea eggs are usually a bit smaller than a chicken egg, more like the size of a bantam chicken, but are almost indistinguishable in taste. The eggs are similar in shape to a turkey egg; more pointed at one end with a wider base at the opposite.

Our guineas are decent layers and lay longer throughout the season than other less-domesticated poultry. Their eggs are also healthier than chicken eggs, because they are higher in vitamins A and B as well as protein, with less bad cholesterol and more good cholesterol.

TERMINOLOGY AND DETAILS

Female bird: Hen

Male bird: Cock

Young bird: Keet

Gestation: 28 days

Egg size: 2 inches long, 4½ inches in circumference at widest point

Egg weight: 2 ounces

If you are interested in raising guineas, start with a couple of birds and see how you like their personalities. Allow them to reach full adulthood before you decide to get more.

Guinea eggs are higher in vitamins A and B as well as protein than chicken eggs.

A baby guinea is called a keet. Guineas are somewhat less domesticated than chickens and do well as free-range birds.

QUAIL

There are more than 130 different breeds of quail, but this section will focus on the Coturnix quail, a very popular breed that is the easiest to locate.

Anyone can raise a covey of quail. They're low-maintenance birds that require very little space. Even apartment dwellers have been known to raise quail on balconies. Most people could keep a small quail coop on their deck without the neighbors ever knowing. Surprisingly, given how easy they are to raise, quail are not as popular among flock enthusiasts as you might think. One reason perhaps is that they're a bit more difficult to locate. Feed stores rarely stock baby quail in the spring. Thus far, the best place to look is by searching for farm swaps, poultry forums, and hatcheries on the Internet.

Quail are raised for both meat and egg production. They are very efficient at converting feed into eggs and meat, with a life cycle substantially quicker than that of a chicken. A Coturnix quail will hatch in 14 to 18 days, and the hens will start laying as early as 6 weeks.

Quail require less space and less feed than chickens do, but they need a higher-protein game bird feed, which can be a bit more expensive to purchase. Even at 24 percent protein, quail feed is still more cost-efficient than feed for chickens or other fowl. The feed should be ground to the size of coarse sand for the small birds to ingest.

The females are very quiet birds, making a small grumble sound. The males make a slightly louder call, shrill in pitch and somewhat similar to the call of a tree frog. The noise is not substantially more than that of a typical songbird, but if noise is an issue in your area, you could keep just one male or raise females exclusively.

Quail eggs are appearing in more and more epicurean markets and higher end restaurants. The quail egg is adorable—a tiny grape-size egg covered in a lovely tan shell with darker brown mottling. The mottling pattern is inconsistent, with a mix of large brown patches and tiny

The quail egg is adorable—a tiny grape-sized egg covered in a lovely tan shell with dark brown mottling.

brown speckles. When hard-boiled, the sweet little eggs can be sliced in half, showing an even smaller yolk. When fried, the white bubbles and crisps up in a lacy pattern, making it a beautiful topper for a small plate dish. In recipes, five quail eggs can be substituted for one large chicken egg. Quail eggs contain more iron than do chicken eggs.

Quail hens are not excellent mothers. In many breeds, the instinct to sit on eggs has been bred out. To continue your flock generation to generation, you'll need to use an incubator or a broody hen, such as a Silkie bantam.

TERMINOLOGY AND DETAILS

Female bird: Hen

Male bird: Cock

Young bird: Chick

Gestation: Coturnix 18 days, other breeds, such as Bob White, 21 to 23 days

Egg size: ¾ inch long, 1 to 1½ inches in circumference at widest point

Egg weight: 1½ ounces

RAISING A MIXED FLOCK

The key to raising a successful mixed flock is timing, observation, having a good plan B (extra holding pens), and lots and lots of space.

Birds can be fickle. Their complex communication patterns and range of personalities can result in animosity even among birds of the same species and breed. The instincts of pecking order, hierarchy, and social status create intricate relationships within a flock, and each time a bird is introduced or removed, the pecking order is reorganized as each individual moves up or down the social scale accordingly.

Social status is complicated further when you mix breeds. In the chicken world, for example, variables such as size, domestication, age, and sex of the breed affect acceptance within the flock. Bantams, though small, can be fierce and may pick on larger breeds. Introducing adult birds to an existing flock can cause intense fighting, depending on the respective personalities and tendencies of the breed.

Social status is complicated even more when you mix species, because each species has a social structure and overall demeanor that is different from the next.

The key to successfully raising mixed species and breeds (even birds of the same breed) is space. As soon as you start noticing pecking, feathers missing, or birds rising up on their haunches or chasing, it indicates you have too many birds in the space provided. Our mixed flock consists of chickens of many breeds and sizes, two breeds of ducks, guineas, turkeys, and geese, and they all coexist harmoniously because they have ample space, plenty of nesting areas, and several feed and water stations. Even with ample space, it's important to have a backup pen whenever you're introducing new birds. Sometimes personalities clash and there is nothing you can do but separate the birds.

Do some research to determine which breeds have the strongest personalities and then keep fewer of those types. For example, guineas can be fierce, so if you raise them, make sure they are outnumbered by other birds.

Another consideration is the age of introduction. In terms of age, there is a sweet spot that makes introducing birds an easier task—the birds need to be old enough to fend for themselves but not yet sexually mature. This is especially important when introducing male chickens. Cockerels should be introduced to adult roosters when they are fully feathered and looking like smaller versions of their adult selves, but before their combs and wattles turn bright red and before they start crowing.

Feeding Mixed Flocks

If you are raising different species for different purposes, you may find that you need to separate the birds so that they are on the appropriate feed. For example, turkeys meant for table should be on a grower/turkey raiser type of feed, which will probably be too rich for other birds, especially laying ducks, which have a tendency to become overweight. You can try to raise feed trays to different levels, but I've found this method unsuccessful because the taller birds will also eat the lower feeds and grow at a slower rate.

Turkeys, Chickens, and Health Concerns

An often-asked question is whether turkeys and chickens can be safely raised together. Unfortunately, there isn't a simple yes-or-no answer to this question.

We have successfully raised adult turkeys and chickens together for years, and their personalities have meshed just fine. However, there is a disease called blackhead that chickens can carry. Although the disease is often unnoticeable in chickens and doesn't affect a chicken's health negatively, it can be deadly to turkeys, especially young poults. Blackhead is usually regional, and you can call your county extension to ask

A Barred Plymouth Rock hen shares a pot of feed with Muscovy ducks.

whether it exists in your area. Even without disease concerns, turkey poults should always be brooded separately from other birds. The poults are extremely delicate and slow, and easily fall prey to more active birds.

The challenge in raising a mixed flock is providing the specific needs of each species in the same housing area. Turkeys need space, geese appreciate grass to graze on, ducks enjoy a body of water, guineas can be territorial, quail are tiny, and chickens encompass a range of all these species' needs. But if you have the space and means to add a new species to your existing flock, do so. Variety is the spice of life—and the backyard coop.

USING EGGS

Over the years, nutritionists have frequently flip-flopped on their views of the health value of eggs. Every ten years or so, the science seems to flip one way or another, and usually the change is centered on the cholesterol levels of eggs.

ALL OF THE CHOLESTEROL found in an egg is contained in the yolk. One large egg yolk contains around 184 milligrams of cholesterol, or 61 percent of your daily value. According to an article by the Mayo Clinic, most healthy adults can safely eat an egg a day without increased risk of heart disease. The cholesterol found in eggs doesn't have the same ill effects as those found in other saturated fats or trans fats.

If you're worried about cholesterol or have a restricted diet, egg whites are a wonderful alternative. They still contain over half the total protein of a whole egg (3.6 grams of the total 6 grams) and contain minimal fat and no cholesterol. Eggs are also high in choline and are a good source of selenium, biotin, vitamin B^{12}, vitamin B^2, vitamin B^5, vitamin D, vitamin A, molybdenum, iodine, and phosphorus.

A while back, the term *omega fatty acids* became very popular in the health food world. Everyone flocked to their local vitamin outlet to get a bottle of fish oil tablets. And for good reason: Omega fatty acids do a number of good things for your body. They promote energy, healthy organ function, strong bones, and good skin, and can help reduce inflammation. Omega-3, omega-6, and omega-9 are considered essential fatty acids because the body does not produce them on their own. You must get them from the food you eat:

- Omega-3 balances the systems of the body.

- Omega-6, although considered an essential fatty acid, is overconsumed by most people, leading to inflammation.

- Omega-9 helps increase good cholesterol and decrease bad cholesterol.

Once your chickens start producing, you're going to be blessed with more delicious, healthy eggs than you know what to do with.

Because we control what types of feeds our chickens consume, backyard chicken keepers can, to an extent, control the quality of the resulting eggs.

Backyard chicken keepers have control over how much free-range our chickens get and what types of feed they consume, and as a result we can, to a certain extent, control how healthful the resulting eggs are. Some feed companies now offer feeds with increased omega fatty acids to the backyard chicken keeper. Conventional egg producers are also responding to the whole foods movement and supplementing their chickens' diet with foods that will increase the amount of omega-3s in the eggs they sell.

Use a whisk to incorporate the white with the yolk and add air to the mix for fluffy eggs. picturepartners/Shutterstock

EGGS IN COOKING

Once your chickens start producing, you're going to be blessed with more delicious, healthy eggs than you know what to do with. And there is something exciting about collecting eggs from your own hens and making a meal from them.

When you begin cooking with eggs, it's important to know how an egg functions as a food item. With some basic understanding, you can create your own recipes and meals from food you already have, perhaps items from your garden.

The egg can be separated into white and yolk, which serve very different functions in recipes.

Eggs as Binder and Barrier

Eggs are one of our most versatile and unique foods—an animal protein we can enjoy without harming the animal. Eggs can be savory or sweet, or added to foods as functional ingredients that don't much change the flavor. Among other things, eggs can be the "glue" used to shape and hold other food items together in the cooking process. For example, eggs are used in savory foods such as meatloaf or meatballs to keep the ingredients from falling apart. They work similarly in foods such as bread pudding, keeping the bits of flavor-soaked bread together.

Eggs also can help batter or bread crumbs adhere to food meant for the fryer or oven. Battering food is usually a three-part process. First, the food is placed in flour, which adds substance to the batter but also dries any excess moisture from the food. Next, the food is dipped in liquid egg to provide a binder coating. Water and oil do not mix, and if you place a piece of wet food in hot oil you will get spattering. Fortunately, eggs do not spatter in oil, and so they make a perfect binder that allows bread crumbs, tempura crumbs, or even cornflakes to adhere. In addition, an egg coating will seal a piece of food as it cooks, creating a barrier that keeps the juices in the food while preventing too much oil from being absorbed.

French toast is a classic example of eggs being used to bind and seal. French toast is essentially bread soaked in a light custard, then fried. Flavors like cinnamon, nutmeg, and sugar are often added, and an egg dip binds these ingredients to the bread, saturating it with flavor. When the bread is pan- or griddle-fried, the egg acts as a barrier, preventing the bread from soaking up too much oil, ensuring a golden, crispy brown exterior and a dense, flavorful center.

Pan-Fried Eggs

Eggs cooked in a frying pan are perhaps the quintessential method of serving eggs. The image of the uneven white disk with the yellow yolk in the center is known to almost everyone. As a kid, I called these "dunk eggs" because my mother always served them with buttered toast to dunk in the yolk. I remember scraping every last golden drop from the center of the egg before I dug into the salty fried white. Most everyone has some individual memory of the role fried eggs played in the breakfast ritual, though the style of cooking may have varied. Here are some basics on pan-cooking eggs, including explanations of the iconic terminology.

- **Over-easy, over-medium, over-hard.** When ordering an egg, or cooking one at home, these labels are much like the preferences for a cooked steak. The term *over* means that the egg will be fried on one side and then flipped—unlike *sunny-side-up*, where only one side of the egg is cooked. The *easy, medium,* and *hard* labels refer to the consistency of the cooked yolk.

  Over-easy means that the egg is flipped and cooked on the second side for only a short time—just long enough for the egg to seal over and keep a delicious pocket of still liquid yolk. This is my favorite way to eat a fried egg.

  Over-medium means the egg is cooked a bit longer on the second side. The egg yolk will begin to thicken.

  Over-hard means the egg is cooked until the yolk is completely cooked through.

  Fried and flipped eggs take a little practice to get right. Eggs are delicate, and it can be difficult to get the whole egg out of the pan and flipped over without folding the white or breaking the yolk. A broken yolk, sadly, means no toast dipping, which is a tragedy at our house.

The success of a flipped, fried egg is in the technique and the proper cooking tools. Although I don't normally like to use nonstick pans, I make an exception when frying an egg, where the nonstick surface, along with a bit of butter, makes the whole cooking process a lot easier. A large, thin flipper also makes it easier to scoop up the complete egg without breaking it. A thin spatula allows you to get under the egg easily. Also, don't overcrowd your pan. Use a larger pan than you need so you'll have some maneuvering room to do the flipping without crashing into other eggs.

1. Put a dab of butter in the pan, about ½ tablespoon per egg. You want a thin layer of butter to coat the pan. I use real butter, not oil or cooking spray. Eggs don't naturally have a strong flavor, and a dab of butter does amazing things to a simple egg. Also, sprinkle salt and pepper on each side.

2. As soon as the butter melts but before it browns, crack your egg into the pan. (You can also crack the egg into a small dish to make sure there aren't any blood spots, broken shells, or other debris.) Let the egg cook over medium to low heat until the white is completely opaque.

3. For *over-easy* eggs, flip as soon as the egg is opaque. For *over-medium* eggs, flip once the edges of the white turn slightly golden and large bubbles start popping through the white. For *over-hard* eggs, turn the heat down a bit and cook longer but slower. Flip once the yolk starts to harden. Then finish cooking thoroughly on the other side.

Sunny-side-up is the easiest pan fried egg to cook. Low and slow will ensure a delicate white and a silky yolk.

Turning an egg is a delicate, practiced dance between pan and flipper. I like to pick up the pan and gently scoot the eggs to one side. Then I use the open space to slide the flipper under the eggs. Next, I scoot the pan back toward the spatula and work the spatula completely under the egg. In one quick sweep, I flip the egg over.

- **Sunny-side-up.** This is perhaps the easiest way to cook a fried egg, as there is no flipping involved. However, many are turned off by the raw appearance of the yolk.

  To cook an egg sunny-side-up, the key is low and slow as far as temperature goes. You want the whites to cook thoroughly (there's nothing worse than slimy egg whites) but the yolk to stay liquid. Keeping the temperature low prevents the whites in contact with the pan from burning before the tops of the whites are cooked solid.

THE FLIP EGG CHEAT

Your mother-in-law is coming for brunch and you haven't mastered your egg flip? Cook the egg sunny-side-up with a lid over the pan. The egg will cook on the bottom and steam over on the top.

Be aware that sunny-side-up and over-easy eggs often don't reach the FDA recommended temperature for cooking an egg, so decide whether you're comfortable with that. This is why some restaurants decline to serve eggs unless they are well cooked.

Poached Eggs

Poaching involves cooking a shell-less egg in hot but not boiling water, at roughly 160° to 180°F. The process intimidates many people, because it is easy to make mistakes. The key is to be patient and gentle. Done correctly, poaching is an ideal way to cook a delicate food like an egg.

1. Fill a medium pot halfway with water. I like to salt the water for a bit of infused flavor. Let the water come up to a simmer. (A rapid boil will disturb the eggs, breaking them into fragments, and you'll end up with cooked egg soup.)

2. Once the water is simmering, crack the egg into a small bowl. Carefully lower the bowl into the simmering water and slide the egg out. Then leave the egg alone. The thin whites will swirl around in the water and may break away from the more solid white area. That's fine. The water may also foam; you can remove any separated egg whites and foam with a slotted spoon.

3. When the egg begins to turn opaque and solidify, take a spoon and gently swish the simmering water over the top of the egg until it is cooked over.

4. Gently scoop the egg from the water with the slotted spoon and let most of the cooking water drip off. You want the egg to dry as best you can before placing it on an English muffin. I've tried using a paper towel, but find that the paper often sticks to the egg. I've found the best thing is a sacrificial slice of bread. Let the egg sit on the bread for a few seconds until it is mostly dry, and then place it on toast or an English muffin. Season with salt and pepper—or better yet, drown in Hollandaise sauce!

Scrambled Eggs

Scrambled eggs are a humble dish, a hearty, no-fuss food. One advantage of scrambled eggs is that you can cook them for the masses, and the technique is the same whether you're making them for 1 or for 21 people.

Scrambled eggs are probably the easiest of eggs to prepare, making them the ideal place to start if you've never cooked an egg before. The keys to delicious, moist scrambled eggs are in the cooking fat and taking care not to overcook them. Scrambled eggs can go from being perfect to overdone in the blink of an eye.

To make great scrambled eggs you will need butter—not margarine or vegetable oil. You can also make them with coconut oil, which doesn't taste of coconut at all and makes surprisingly good eggs. You will also need cream. Full-fat milk is acceptable, but cream does magical things to scrambled eggs. The liquid lends moisture to the eggs while cooking, and the fat helps seal in the moisture.

Choose a pan size appropriate to the quantity you're making—large enough to allow you to move the eggs around but not so large that the eggs will spread out and cook too quickly. A 6- to 7-inch pan for two eggs works well. Nonstick will make your life easier, but if using a regular pan, add a little extra butter.

1. Crack your eggs into a bowl. I usually figure two eggs per person for an average serving size. Add a dash of cream and whisk them well. The more you whisk, the more air will whip into your eggs, making them fluffier.

2. Add the butter to a cold pan and warm over medium heat until it's almost melted. You don't want to brown anything.

3. Add the eggs and immediately start moving them around in the pan. When cooking scrambled eggs, the eggs should constantly be on the move. To create this movement, my

(above) Scrambled eggs are probably the easiest of eggs to prepare, making them the ideal place to start if you've never cooked an egg before.

(right) A flat-bottomed wooden spatula is the perfect tool for making scrambled eggs.

favorite utensil is the flat-bottomed wooden spatula. In a clock-like fashion, scrape the pan from the outside inward, piling the eggs in the center, then spreading them out and starting again. If the eggs start to stick to the bottom, ignore the stuck egg and leave it on the bottom of the pan and move the loose eggs on top.

4. To avoid overcooking the eggs, remove from the heat before they are done all the way and let them finish cooking from the residual heat still in the pan. Serve immediately.

Scrambled eggs can be delicious on their own, but they also act as a base for many different flavor combinations. Try adding vegetables, breakfast meats, and a variety of cheeses. One tip: Scrambled eggs take only a few minutes to cook. So if you're adding raw ingredients, those foods might still be raw by the time your eggs are done. Be sure to cook flavor additions first, and then add them to the eggs. Scrambled eggs make a great base for leftovers.

Omelets

The omelet is like the scrambled egg's stuck-up cousin. While scrambled eggs are like a heap of wrinkled laundry, the omelet is a neatly folded garment.

Omelets, like scrambled eggs, are often used as a carrier for other cooked ingredients or cheese, but in this case those ingredients are folded inside a pocket of cooked egg, not all mixed up.

Omelets present a few challenges when cooking. Again, the nonstick pan is almost a requirement for the fussy omelet. You will also need a good layer of fat because, unlike scrambled eggs, where some sticking is acceptable, the omelet needs to slip out completely.

1. A traditional portion size is a three-egg omelet. You start in the same way you do with scrambles eggs—whisking the eggs until uniform in color and texture. Add a bit of salt and pepper.

2. Melt the butter in the pan over medium to low heat and pour in the eggs. For the first few seconds you can move the eggs around with a spatula. As the eggs separate from the pan, tilt the pan so that more liquid egg can fill in that space. Coat the bottom of the pan completely with eggs and then leave them be.

3. When the eggs begin to cook on the bottom but before they are solid on top, add your filling ingredients. Keep the heat low, and watch for the tops of the eggs to become firm. As soon as this happens, it's time to fold.

4. Loosen the eggs from around the pan and slide them to one side of the pan. Get the spatula under the eggs and fold one side over the other, enveloping the filling ingredients.

Dropping their temperature quickly will make it easy to peel fresh, hard-boiled eggs.

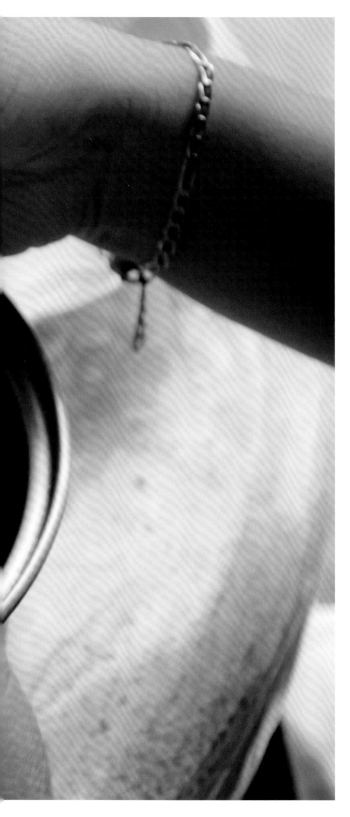

Hard-Boiled Eggs

Keeping a bowl of hard-boiled eggs in the refrigerator is a great way to have protein-rich snacks ready to grab on the go. Whenever I make hard-boiled eggs, I always boil a few extra to have on hand. It makes breakfast a snap, and provides ingredients for salads or egg salad sandwiches.

Fresh eggs that are hard-boiled are notoriously hard to peel. This is because a freshly laid egg is full and plump inside its shell, and then begins to lose volume as it ages. When a fresh egg is boiled, it expands from the heat and presses tight against the shell, where it sticks. The result can be a mess when you try to peel it after boiling. When cooking fresh eggs, you might notice that the eggs expand so much that the shell cracks slightly in the water. An older egg, though, shrinks in size, so the contents do not stick to the inside of the shell with the same vigor, making it easier to peel.

Remember that heat expands, and cold retracts. Cooling the eggs as quickly as possible after cooking will shrink the hot egg back into its shell and make it easier to peel. I also find that peeling an egg under cold running water helps lubricate the shell and makes it easier to slide the shell off.

1. Fill a pot with cold water and add the eggs. I also add a dash of salt, as this increases the boiling temperature. Turn on the heat to high. Once the water starts to boil rapidly, set the timer for 9 minutes. While the eggs are boiling, fill a bowl with ice water.

2. When 9 minutes have passed, immediately take the eggs to the sink. Pour out the hot water and run cold tap water over the eggs.

3. Drain the water again and place the eggs into the bowl of ice water while you're peeling. The faster you can drop the temperature after boiling, the more the egg will shrink and the easier it will be to peel.

MAKE YOUR PASTRIES GLISTEN

To make pie crusts and pastries shiny and golden brown, whisk an egg with a bit of milk and brush on the crust before baking. This mixture can also act as a glue to keep a dusting of coarse sugar in place on scones and turnovers.

In the kitchen, you'll find that store-bought eggs pale in comparison (literally) to those produced by your own poultry.

EGGS IN BAKING

Eggs are an essential ingredient in almost all types of baking, where they serve several functions. Eggs create rise in baked goods; they also create denseness and can help bind air into a recipe. The yolk fats help emulsify other ingredients and the whites lend strength to the finished shape. Eggs also tend to dry a recipe, so they must be counterbalanced with wet or oil-based ingredients to keep moistness in bake goods.

METHODS FOR PRESERVING EGGS

Spring is prime egg-laying season. As fall approaches and the daylight hours become shorter, chickens begin their winter break. Birds are very sensitive to the number of daylight hours; shorter daylight periods not only signals them to stop laying but also prompts many migratory birds to start heading south for the winter. You can "trick" your chickens into laying more eggs by supplying an artificial light source in the coop, but your production will never match the high yield of the spring. So if you want to enjoy plentiful quantities of eggs year-round, some form of preserving will need to be part of your skill set.

Luckily, eggs can be stored for periods of time when production is high. None of these methods will match a fresh-cracked egg, but it will get you through the lighter months without having to buy eggs at the grocery store.

Eggs can be frozen in cupcake trays for use in winter baked goods.

Freezing

One of the easiest long-term storage methods is freezing eggs. Eggs should not be frozen in the shell, as the liquid egg will expand and crack the shell open in the freezing process. Instead, crack the eggs into cupcake trays or ice cube trays and freeze up to 6 months. I like to freeze them individually, then once they are hard I transfer them to zip-top food storage bags.

Frozen eggs are not recommended for frying, but work well for scrambled eggs or for use in baked goods. Eggs that have been frozen should be cooked thoroughly. To use, thaw in the refrigerator overnight.

I don't recommend freezing whole hard-boiled eggs, as the whites will break down and become watery, but egg yolks freeze very well. Lay hard-cooked yolks on a sheet tray in a single layer. When frozen, place them in a bag for easier storage.

Preserved hard-boiled eggs.

Curing

This process is so interesting to me, and the flavor is so unique that it's almost like eggs have become a new food. Cured egg yolks are like a flavorful garnish that, when grated over food, almost act like a hard cheese. You get a dense, salty flavor that is delicious. It is an excellent garnish for pastas. There are a variety of recipes available for curing egg yolks, but the essence of the process could not be simpler.

1. First, separate the egg yolks from the whites and embed them in a blanket of salt and sugar (or salt alone, in some variations).

2. Let the yolks cure in the salt/sugar mix in the refrigerator for about a week.

3. Uncover the yolks, which by now are hardened, and let them continue to dry in the refrigerator, covered in cheesecloth, for another week or two. The cured and hardened egg yolks will now keep for a month or so in an airtight container, and can be grated over foods as a garnish.

   In some variations of this process, the yolks are dried in an oven set at a low temperature rather than air-dried in the refrigerator.

OTHER USES FOR EGGS AND EGGSHELLS

Raising chickens will provide you with an abundance of eggs in the spring and summer—and also an abundance of eggshells. Here are some ideas for using those:

- Eggshells are wonderful to use in the garden. They can be added to compost or ground up and applied to soil. Tomatoes especially benefit from the calcium that eggshells provide.

- Seedlings can be planted directly in eggshell halves filled with starting soil.

- Plants are not the only ones that benefit from eggs in the backyard homestead. Chickens can be fed dried, ground eggshells in lieu of store-bought calcium supplements like oyster shell. To use, rinse eggshells and allow to air-dry (or you can speed the process in a low oven). When dry, crush into small pieces and feed free choice in a separate feeder from the main feed.

Eggshells make perfect growing containers for seedlings. They provide nutrients to the plant as they decompose and can be planted right in the soil without disturbing the roots.

CAN EGGS MAKE YOU BEAUTIFUL?

Although eggs provide many nutrients that can promote a healthy body and a radiant glow, they can also be used on the outside for minor skin issues. For example, egg whites can be applied to the skin and allowed to dry to help shrink pores, draw out impurities, and combat oily skin.

Egg yolks can also help balance dry skin. Whisk together an egg yolk, a drizzle of olive oil, and a dash of cream. Eggs have emollient properties that help soften the outer layers of skin, making it easier for the skin to absorb the oil. The lactic acid in the cream helps slough off dead skin cells.

If you simply have too many eggs to use or give away, remember that chickens can also be fed the whites and yolks of eggs as a nutritional supplement. This is especially beneficial during stressful times, such as the cold winter months or during molt. Just be sure to cook the eggs to prevent self-appointed egg eating in the coop, because once chickens learn that raw eggs are delicious, they'll take it upon themselves to crack open laid eggs and eat away.

TROUBLESHOOTING

When you grow up eating store-bought eggs, you get used to eggs looking a certain way—for the most part, they're white, clean, and uniform in shape and size. When you raise chickens of your own, however, you soon realize that all eggs are not formed in a cookie-cutter mold. Each chicken is different, and while most eggs look relatively similar, you will occasionally get an odd egg.

An unusually large egg may be a double-yolk egg.

THE EGGS YOU FIND in your nest boxes provide an external clue as to what's going on inside your hen. If she's laying normal, healthy eggs on a regular basis, chances are your hen is also healthy on the inside. If something is out of whack with your hens, egg laying is one of the first things affected. In this chapter, I will address some common oddities in eggs and why your hens may stop laying altogether.

ODD EGGS AND YOUNG LAYERS

Often, the strangest eggs will come from young layers. A young hen is new to the egg-laying process. Her system is still working itself out and it will take time to fine-tune her egg production. Don't fret if you reach into your nest box and pull out something weird. Most of the time this is normal and not cause for concern. Many of these problems will pass with time; she simply needs to lay a few eggs and mature to a point where her eggs become normal.

Wind Eggs

A wind egg is a small, yolkless egg. It occurs when a small piece of reproductive tissue breaks free and the hen's system mistakes it for a yolk. The fragment moves down the oviduct and the egg is formed around the imposter yolk.

If your hen is laying normal, healthy eggs on a regular basis, chances are she is also healthy on the inside.

Large Eggs and Double Yolks

A double yolk egg is formed when the hen releases two yolks at once. The twin yolks travel down the oviduct and the egg is formed around both yolks, often resulting in an extra-extra-large egg. Double yolks are common among young pullets that are still getting their egg-laying system in working order. In other instances, the double-yolk tendency is genetic. You may find that a hen will consistently lay double-yolk eggs throughout her life.

Blood on the Shell

The most common cause for a blood smear or spot on the shell is, again, immaturity in a young hen. Her duct has not stretched completely, so blood can appear on the shell until her system becomes more accustomed to laying.

An Egg within an Egg

It's a rare occurrence but an interesting one nonetheless. A hen can lay an egg within an egg if she releases a second yolk before the first egg is laid. This release causes a contraction that reverses the first egg being laid and sends it backward up the oviduct. As the second egg comes down, it encapsulates the first egg in the shell-making process and causes an egg within an egg. The final egg is usually extra large.

Oddly Shaped Eggs

Usually, an odd-shaped egg is a fluke occurrence and no cause for concern. It's especially common in both young and old hens. A young hen, as we've discussed, may lay strange eggs as her oviduct is developing. An older hen may lay odd-shaped eggs as her reproductive system begins shutting down. If a hen in her egg-laying prime lays an occasional misshapen egg, this could be due to stress or disruption while the shell was forming. If she consistently lays a misshapen egg, it may be time to talk to a veterinarian about possible respiratory diseases, such as Newcastle, or a problem with the hen's oviduct.

Wrinkled Eggshell

A wrinkle can appear in the eggshell if the hen was disturbed, bumped, or stressed while forming the egg. This is often referred to as a "body check egg." The wrinkle occurs as the oviduct attempts to heal the disruption in the egg-formation process, similar to the way scar tissue forms.

Shell-less Egg or Soft Egg

A shell-less egg is an egg that appears rubbery, lacking the hard outer shell. The most common reason for this is a calcium deficiency. If you are getting shell-less eggs, make sure that your chickens are on a layer formula, and offer them a free-choice calcium supplement in the form of ground eggshells or ground oyster shell.

Triple yolks are not unheard of!

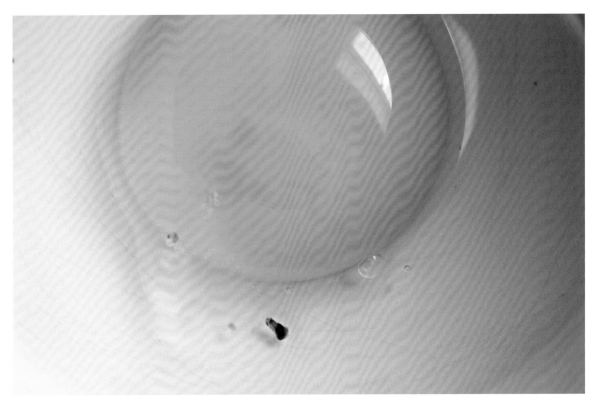

You might find that your farm-fresh eggs have more blood spots than store-bought eggs. The truth is that these blood spots occur just as often in factory-farmed hens.

Rough-Texture Shell

Some eggs will have chunks of raised shell matter along the surface that give the shell a bumpy or rough texture. This is caused by excess calcium. Make sure that any calcium supplements are offered free choice.

Lash Eggs

Lash eggs are not eggs at all, though they may have some egglike properties. Instead, they are the result of a bacterial infection that causes salpingitis, or an inflamed oviduct. The inflamed duct releases pus, mucus, and tissue that travel down the duct. The imposter material is treated like a yolk, resulting in a misshapen shell and some egg material. If you find a lash egg you should contact your veterinarian immediately.

Blood Spots

A blood spot is a small imperfection inside the egg that appears as a dark brown or reddish tissue usually near the yolk. It occurs when a blood vessel ruptures on the surface of the yolk or in the oviduct and is then trapped in the egg as it forms. While unsightly, eggs with blood spots are still perfectly safe to eat when cooked to temperature.

You might find that your farm-fresh eggs have more blood spots than store-bought eggs. The truth is that these blood spots occur just as often in factory-farmed hens, but large-scale production companies often candle their eggs to check for blood spots and those eggs are removed before packaging.

WHY DO CHICKENS STOP LAYING?

The purpose of laying eggs, as far as a chicken is concerned, is to further the population of the flock. Because of this, nature has designed a chicken's body to stop laying in times of stress. If a chicken is experiencing stress, nature knows that it's not a good time for her to become a mother. Halting egg production also allows the chicken to put energy that would otherwise be used to produce eggs into healing her body.

Stress can be natural and even seasonal. A break from laying eggs allows the chicken to recuperate from her last cycle and prepare for the next. This fact may frustrate us as chicken keepers, as we dump pounds of feed into their feeder and come out with an empty egg basket. But remember that the formation of an egg takes a lot from a chicken. She sacrifices protein, calcium, and calories from her own body to produce an egg every day.

With all the hard work that it takes to lay an egg, the absence of eggs is a good indicator that your chickens are experiencing a change. Dry periods are often a normal part of a chicken's life cycle, but they can also hint that something is off with your chicken's health.

Normal Causes for Empty Nest Boxes

Age Each breed of chicken starts laying at a different age. On average, a chicken will start laying her first eggs around 6 months of age. As a chicken ages, she will lay steadily for about 3 years, and then lay fewer and fewer eggs each year thereafter. Some of our old hens lay an egg once a month during the warmer months.

Molt Usually in the late summer or early fall of a chicken's second year, it will experience its first molt—the natural and healthy process by which chickens replace their old feathers with new ones. Molt is triggered by the waning number of daylight hours as the northern hemisphere tilts away from the sun, and is naturally timed so that a chicken has the best, most perfect plumage as it enters the cold winter months.

Different chickens experience molt in different ways. With some, it's hardly noticeable as a few feathers drop here and there. Others molt heavily, becoming bald and straggly.

Growing feathers back takes lots of extra energy, and because feathers are mostly made of protein, laying stops during the molt to divert protein into feather production. A molt usually lasts between 8 and 12 weeks, but you can speed this period by providing your chickens with additional protein in their feed.

Molt can also be brought on by other things besides the change in seasons. Stress, change in diet, travel, or a change in pecking order can bring on a molt that halts egg production.

Shorter daylight hours Eggs are a seasonal product. Just as tomatoes ripen in August and pumpkins are picked in October, eggs are mostly laid in spring and summer. Not only does the limited daylight of fall bring on seasonal molt, but it is also responsible for a seasonal break from egg laying. If your hens molt late in the fall, you may not see eggs again until the following spring. Although 6 months is the normal age to begin laying, a first-year pullet that hatched too late in the spring may not be exposed to enough daylight hours to stimulate egg production. In this case, the hen may be egg-free for the first year.

Biologically, it makes sense that a chicken would stop laying eggs in the cold, short winter days, as nature intends for the mother hen to keep her chicks a toasty 95°F. For this reason, most chickens will slow egg production in the winter months and some stop altogether. A hen's body is using energy to keep warm, to recuperate from the fall molt, and to recover from the egg-laying months. Diet is also less varied in the winter, when there are fewer insects and less vegetation to feed on.

In the backyard setting, however, we can supplement what nature fails to provide in the winter months with healthy, balanced feed. Because of this, some people decide to "trick" their chickens into laying by providing artificial light in the coop. Chickens need approximately 14 hours of daylight to trigger egg production. If artificial light is provided, it's better to add it in the morning, so that a normal, gradual sunset period triggers their instinct to return to the coop to roost. The easiest way to control the amount of light in your coop is to set up a timer system.

The broody hen   Another cause of decreased egg laying is broodiness. "Broody" is the term used to describe a hen that has developed the instinct to be a mother. Some breeds are more prone to broodiness than others. During this time, a hen will lay a clutch of eggs or claim another chicken's eggs. When she has a sufficient number, she stops laying to concentrate on incubating the clutch. She also plucks the feathers from her under-breast to provide skin-to-egg contact, and will run a slight temperature as her body goes into a protective trance. She leaves the nest only once a day to defecate, eat, and drink, and becomes incredibly protective and even aggressive should anyone come near her precious clutch.

If you keep a rooster with your hens, then this is a great way to increase your flock without an incubator. If you want to increase your flock, then by all means allow the hen to hatch her chicks. But be aware that hens will sit on eggs whether they are fertile or not. Allowing your hen to fruitlessly sit on eggs will take a toll on her body.

It can be very difficult to break a hen of her broodiness. One of the best ways to prevent broodiness is to be diligent about collecting eggs. As soon as eggs begin to pile up, collect them to remove the temptation to brood. Still, some hens have such a strong instinct to brood that they will do so without even having eggs to sit on.

Kicking the hen out of the nesting box as often as possible is somewhat effective, but sometimes a hen needs a more drastic change. I find that placing her in new surroundings for an extended period is most effective. Lock her outside in a run or in a free-range setting away from her selected nest. Usually a day or two is enough to break the trance, but some really focused hens might need more time.

Egg-bound hen   When a chicken is "egg bound" it means exactly that—the egg has become lodged between the uterus and the cloaca. Often, this is due to a large egg or a double yolk. When a hen is laying an egg, a section of the intestines closes off, and when she becomes egg bound, she's unable to have a bowel movement and can die quickly. This is more than just a "no eggs" problem, it's one that can kill your hen.

Parasites and illness   Serious health problems can, of course, cause a hen to stop laying. Fighting infection or a parasite infestation can take a toll on a hen's health, and she will sacrifice her egg laying to heal her body.

Dehydration   Eggs are made of 75 percent water, and if a hen becomes dehydrated she cannot form eggs. Chickens should always have access to fresh, clean water. Chickens will often stop laying in extreme heat to conserve the water in their systems. In cold climates, take steps to ensure that water does not freeze.

Predators   Sometimes a lack of eggs is not the hen's doing. Many predators, such as snakes and skunks, will steal eggs. Make sure that your chicken coop is predator-proof.

Broody hens are one common cause of drops in egg production. Broody hens stop laying to concentrate on incubating their clutch.

INTERESTED IN SELLING YOUR EGGS?

If you're interested in selling your extra eggs, there are three questions to ask: Is it legal to sell eggs in my area? Is there a market for my fresh eggs? Can I make a profit?

Start with your local county extension, township, or city offices, which can help decipher the laws in your area. Topics to ask about:

- Are there required vaccinations for my flock?

- Are inspections or health screenings required?

- Are there washing/cleaning and packaging requirements for the eggs?

- Are food and/or business licenses required?

- Are business signs allowed to be displayed, such as "eggs for sale"?

Next, carefully consider whether there is even a demand for your eggs. When you live in a rural area, for example, all your neighbors may have chickens and already have more eggs than they can eat. Don't go down this path if there are no nearby customers for your eggs.

Finally, determine whether the likely profits really justify the costs of keeping chickens or increasing your flock. There's really no problem if you're already raising chickens for pleasure and simply want to make a few extra dollars from eggs you can't eat yourself. But as a true moneymaking venture, there are many cost considerations, such as the coop, feed, bedding, electricity, cartons, advertising, and the birds themselves.

If you're raising chickens as a serious business venture, you want stellar layers. Leghorns are a great breed for this purpose. You also want chickens that are hardy for your area. Wyandottes do great in the cold and are reliable layers.

When people buy farm-fresh eggs, they may be intrigued by those with colored shells. Green, blue, and dark brown are always a hit. Consider adding Easter Eggers, Olive Eggers, Crème Legbars, or Marans to your flock.

Another thing to consider is the appeal of eggs from other poultry. Duck eggs are becoming very popular, as are quail and guinea eggs. Unusual eggs may command a higher price per dozen.

Dehydrated hens cannot form healthy eggs. Be sure your flock always has access to clean water in all weather.

SOURCES

CHAPTER 2

"The Anatomy of a Chicken Egg."
www.imaginationstationtoledo.org/
educator/activities/how-to-make-a-naked-egg/
the-anatomy-of-a-chicken-egg

"Anatomy of an Egg."
www.exploratorium.edu/cooking/eggs/
eggcomposition.html

Dickinson, Katherine. "Six Things You Didn't Know
About Chicken Reproduction," June 9, 2012.
www.realclearscience.com/blog/2012/06/
chicken-reproduction.html

"Formation of the Egg."
www.thepoultrysite.com/publications/1/
egg-quality-handbook/2/formation-of-the-egg/

CHAPTER 3

Alterman, Tabitha. "Eggciting News !!!," *Mother Earth
News*, October 15, 2008.
www.motherearthnews.com/real-food/
pastured-eggs-vitamin-d-content.aspx

Clauer, Phillip J. "Proper Handling of Eggs: From Hen
to Consumption," Virginia Cooperative Extension.
www.pubs.ext.vt.edu/2902/2902-1091/2902-1091.
html

"How to Grade and Size Eggs."
www.southernstates.com/articles/size-grade-eggs.
aspx

Jacob, Dr. Jacquie. "Feeding Chickens for Egg
Production," University of Kentucky,
www.articles.extension.org/pages/69065/
feeding-chickens-for-egg-production

CHAPTER 4

Adkerson, Tim. "A Review of Egg Color in
Chickens." www.maranschickenclubusa.com/files/
eggreview.pdf

CHAPTER 5

Buchholz, Richard. "Raising Coturnix Quail,"
Mother Earth News, September/October 1981.
www.motherearthnews.com/homesteading-
and-livestock/coturnix-quail-
zmaz81sozraw?PageId=5

CHAPTER 6

Egg Nutrition Center. "Eggs 101 Egg Nutrition
Facts." www.eggnutritioncenter.org/egg-101/

Lopez-Jimenez, Francisco MD. "Are chicken eggs
good or bad for my cholesterol?" Mayo Clinic
www.mayoclinic.org/diseases-conditions/
high-blood-cholesterol/expert-answers/cholesterol/
faq-20058468

"Omega-3, -6, -9 — Making Them Count For You."
www.omega-9oils.com/omega-9-advantage/
healthier-profile/omega-3-6-9.htm

Gunnars, Kris BSc. "Pasture vs Omega 3 vs
Conventional Eggs: What's the Difference?"
www.authoritynutrition.com/pastured-vs-
omega-3-vs-conventional-eggs/

"The Chicken Egg Page," *Mother Earth News*.
www.motherearthnews.com/homesteading-
and-livestock/eggs-zl0z0703zswa.aspx

CHAPTER 7

Zadina, Chad and Sheila E. Scheideler. "Proper Light
Management for Your Home Laying Flock,"
University of Nebraska–Lincoln.
www.hort.purdue.edu/tristate_organic/
poultry_2007/Light%20Management.pdf

"Egg Binding in Hens," July 31, 2014.
www.backyardchickens.com/a/egg-binding-
symptoms-treatment-and-prevention

PHOTO CREDITS

Author illustrations:

pp40–41, pp46–47, p74.

Shutterstock:

pp8–9, Kornnisa; p10, lightpoet; p22, Fotosr52; p29, Sherjaca (upper right); p42, Einar Muoni; p49, Anneka (right); p77, Frank Chen Photography; p83, gengirl; p101, Jiang Hongyan; p104, jadimages (top); p111, Andrey Tiyk; p125, TreesG; p128, Malivan_luliia; p130, Ksu Shachmeister; p131, picturepartners; p135, Bildagentur Zoonar GmbH; p136, Ivan Dragiev; p138, Anirut Thailand.

Emma Squire: p48

INDEX

Page numbers in *italics* refer to figures.